FOOD SCIENCE AND TECHNOLOGY

FOOD SAFETY

CONTAMINANTS, PATHOGENS, AND ILLNESSES

FOOD SCIENCE AND TECHNOLOGY

FOOD SCIENCE AND TECHNOLOGY

FOOD SAFETY

CONTAMINANTS, PATHOGENS, AND ILLNESSES

GEORGES MARTIN
EDITOR

snova
New York

NOTICE TO THE READER

Library of Congress Cataloging-in-Publication Data

ISBN: 978-1-53616-762-7

Published by Nova Science Publishers, Inc. † New York

CONTENTS

PREFACE

The Food and Drug Administration (FDA) is responsible for overseeing the safety of about 80 percent of the nation's food supply and for promoting good nutrition. Chapter 1 reviews FDA's food safety- and nutrition-related activities and resources.

Arsenic, an element in the earth's crust, can be harmful to human health and may be present in water and certain foods, such as rice. Chapter 2 examines what the National Research Council (NRC) and recent key scientific reviews have reported about the effects of ingestion of arsenic on human health, the extent to which FDA and USDA have managed the risk to human health from arsenic in rice, and the extent to which FDA has coordinated with USDA and other federal agencies on actions to manage the risk.

Most seafood consumed in the United States is imported, and about half of it is raised on fish farms. Because farmed seafood is raised in confined areas and susceptible to infections, farmers may use drugs like antibiotics. The use of unapproved drugs or the misuse of approved drugs may result in unsafe residues in seafood that can cause cancer or allergic reactions, according to FDA, which is charged with ensuring the safety of most seafood. Chapter 3examines how FDA helps ensure the safety of imported seafood from unsafe drug residues and ways the agency could strengthen its efforts; how FSIS helps ensure the safety of imported catfish from unsafe drug residues and ways the agency could strengthen its efforts; and the extent to which FDA and FSIS coordinate their oversight efforts.

The U.S. food supply is generally considered safe, but the Centers for Disease Control and Prevention (CDC) estimate that Salmonella and Campylobacter in food cause about 2 million human illnesses per year in the United States. Chapter 4 reviews USDA's approach to reducing pathogens in meat and poultry products. The produce industry continues to be associated with a series of foodborne illness outbreaks across multiple U.S. states and Canada, resulting in hundreds of illnesses and hospitalizations, as well as kidney failure and death in some cases as reported in chapter 5.

The Trump Administration is proposing to consolidate the federal government's primary food safety functions into a single federal agency based in the U.S. Department of Agriculture (USDA) as reported in chapter 6.

Chapter 1 - FDA is responsible for overseeing the safety of about 80 percent of the nation's food supply and for promoting good nutrition. Federal oversight of food safety has been on GAO's high- risk list since 2007. According to FDA, FSMA aims to improve food safety by shifting FDA's focus toward preventing food contamination, rather than responding to foodborne illnesses. GAO was asked to review FDA's food safety- and nutrition-related activities and resources. This chapter examines (1) FDA's key food safety- and nutrition-related activities since FSMA's enactment in 2011 and how FDA determined its priorities for those activities, (2) the resources FDA dedicated to those activities in fiscal years 2011 through 2016, (3) the extent to which FDA set goals for those activities in fiscal years 2011 through 2017 and is assessing progress toward those goals, and (4) FDA's planned food safety- and nutrition-related activities and associated time frames. GAO analyzed FDA documents and data for fiscal years 2011 through 2018 and interviewed FDA officials.

Chapter 2 - Arsenic, an element in the earth's crust, can be harmful to human health and may be present in water and certain foods, such as rice. Rice may be more susceptible to arsenic contamination than other crops due to the flooded conditions in which it is typically grown. FDA and USDA work to address food safety risks. FDA's responsibilities for rice include regulatory and research programs; USDA's include research programs. GAO was asked to review issues related to arsenic and rice. GAO examined (1) what NRC and recent key scientific reviews have reported about the effects of ingestion of arsenic on human health, (2) the extent to which FDA and USDA have managed the risk to human health from arsenic in rice, and (3) the extent to which FDA has coordinated with USDA and other federal agencies on actions to manage the risk. GAO analyzed a 2013 NRC report on inorganic arsenic, 14 reviews of scientific studies on the human health effects of ingesting arsenic published from January 2015 to June 2017, and agency documents; interviewed agency officials; and compared good practices with actions FDA and USDA took to manage risk and that FDA took to coordinate.

Chapter 3 - Most seafood consumed in the United States is imported, and about half of it is raised on fish farms. Because farmed seafood is raised in confined areas and susceptible to infections, farmers may use drugs like antibiotics. The use of unapproved drugs or the misuse of approved drugs may result in unsafe residues in seafood that can cause cancer or allergic reactions, according to FDA, which is charged with ensuring the safety of most seafood. Beginning in April 2016, FSIS became responsible for ensuring the safety of imported catfish. This chapter examines (1) how FDA helps ensure the safety of imported seafood from unsafe drug residues and ways the agency could strengthen its efforts; (2) how FSIS helps ensure the safety of imported catfish from unsafe drug residues and ways the agency could strengthen its efforts; and (3) the extent to which FDA and FSIS coordinate their oversight efforts. GAO reviewed information from each agency and interviewed agency officials and other key stakeholders.

Chapter 4 - The U.S. food supply is generally considered safe, but the Centers for Disease Control and Prevention (CDC) estimate that *Salmonella* and *Campylobacter* in food cause about 2 million human illnesses per year in the United States. In 2014, GAO identified challenges USDA faced in reducing pathogens in poultry products, including standards that were outdated or nonexistent and limited control over factors that affect pathogen contamination outside of meat and poultry slaughter and processing plants, such as practices on the farm. GAO was asked to review USDA's approach to reducing pathogens in meat and poultry products. This chapter examines (1) the extent to which USDA has developed standards for meat and poultry products and (2) any additional steps USDA has taken to address challenges GAO identified in 2014. GAO reviewed relevant regulations, documents, and data and interviewed officials from USDA and CDC, as well as 17 stakeholders representing industry, consumer groups, and researchers selected based on their knowledge of USDA's meat and poultry slaughter inspections and food safety.

Chapter 5 - The produce industry continues to be associated with a series of foodborne illness outbreaks across multiple U.S. states and Canada, resulting in hundreds of illnesses and hospitalizations, as well as kidney failure and death in some cases. Many in Congress have expressed concern that foodborne illness outbreaks are continuing to occur despite ongoing regulatory oversight by the Food and Drug Administration (FDA) and state public health authorities and given the enhanced authorities and resources provided to food safety officials following the enactment of comprehensive food safety legislation as part of the FDA Food Safety Modernization Act in 2010 (FSMA, P.L. 111- 353). FSMA req;lllloluired the development and implementation of produce safety standards for certain fruits and vegetables, among other regulatory actions.

In: Food Safety
Editor: Georges Martin
ISBN: 978-1-53616-762-7

Chapter 1

FOOD SAFETY AND NUTRITION: FDA CAN BUILD ON EXISTING EFFORTS TO MEASURE PROGRESS AND IMPLEMENT KEY ACTIVITIES*

United States Government Accountability Office

ABBREVIATIONS

Congressional Budget Justification	Justification of Estimates for Appropriations Committees
FDA	Food and Drug Administration
FFDCA	Federal Food, Drug, and Cosmetic Act
FSMA	FDA Food Safety Modernization Act
FTE	full-time equivalent

---

* This is an edited, reformatted and augmented version of United States Government Accountability Office; Report to Congressional Requesters, Accessible Version, Publication No. GAO-18-174, dated January 31, 2018.

FVM	Foods and Veterinary Medicine
GPRA	Government Performance and Results Act
GPRAMA	GPRA Modernization Act of 2010
OMB	Office of Management and Budget
Unified Agenda	Unified Agenda of Federal Regulatory and Deregulatory Actions
USDA	U.S. Department of Agriculture

WHY GAO DID THIS STUDY

FDA is responsible for overseeing the safety of about 80 percent of the nation's food supply and for promoting good nutrition. Federal oversight of food safety has been on GAO's high- risk list since 2007. According to FDA, FSMA aims to improve food safety by shifting FDA's focus toward preventing food contamination, rather than responding to foodborne illnesses.

GAO was asked to review FDA's food safety- and nutrition-related activities and resources. This chapter examines (1) FDA's key food safety- and nutrition-related activities since FSMA's enactment in 2011 and how FDA determined its priorities for those activities, (2) the resources FDA dedicated to those activities in fiscal years 2011 through 2016, (3) the extent to which FDA set goals for those activities in fiscal years 2011 through 2017 and is assessing progress toward those goals, and (4) FDA's planned food safety- and nutrition-related activities and associated time frames. GAO analyzed FDA documents and data for fiscal years 2011 through 2018 and interviewed FDA officials.

WHAT GAO RECOMMENDS

GAO is making three recommendations, including that FDA (1) develop performance measures with associated targets and time frames for all eight of its food safety- and nutrition-related objectives and (2) complete a plan that includes specific actions, priorities, and milestones for implementing the FVM Program's strategic plan. The agency agreed with GAO's recommendations and identified actions to implement them.

WHAT GAO FOUND

From the enactment of the FDA Food Safety Modernization Act (FSMA) in January 2011 through September 2017, the Food and Drug Administration (FDA) has conducted numerous food safety- and nutrition-related activities, determining its priorities for those activities based on statutes and its strategic goals. More specifically, FDA published 33 proposed or final key regulations and 111 draft or final key guidance documents, focused mainly on food safety. FDA also conducted other key activities related to food safety and nutrition, such as conducting inspections and developing risk-assessment tools, responding to foodborne illness outbreaks, and providing outreach and education.

FDA dedicated at least $1 billion annually, including salaries for at least 4,300 full-time equivalent (FTE) staff, to food safety and nutrition activities in fiscal years 2011 through 2016. About 98 percent of those resources were dedicated to food safety each fiscal year, and about 2 percent were dedicated to nutrition.

Since fiscal year 2011, FDA has set goals for its food safety- and nutrition- related activities but has not fully developed the necessary framework to assess progress toward those goals. Most recently, for fiscal years 2016 through 2025, the agency's Foods and Veterinary Medicine (FVM) Program, which is primarily responsible for carrying out these activities, has set a food safety goal to protect American consumers from foreseeable hazards and a nutrition goal to foster an environment that promotes healthy and safe food choices. These goals are supported by eight strategic objectives. The program has developed performance measures to assess progress toward five of these objectives but not for the other three. For each developed performance measure, FDA reports both targets set and measurements taken for specific time frames. For one such measure related to FDA's evaluation of food safety hazards, FDA targeted the completion of 50 percent of evaluations by their due dates in the first quarter of fiscal year 2017 and achieved 89 percent. Leading practices in performance management state that federal programs should use performance information to achieve program goals, and each objective should be tracked through performance measures that have targets and time frames. According to agency officials, the program is developing additional measures for its food safety- and nutrition-related objectives, but it had not finalized them as of January 2018. Until the program develops measures with associated targets and time frames for all eight objectives, FDA cannot fully assess progress toward achieving its goals.

FDA has identified food safety- and nutrition-related activities that it plans to undertake in fiscal year 2018, but its time frames for such activities in the longer term are unclear. According to FDA officials, the agency plans to pursue the food safety and nutrition strategies identified in the FVM Program's 10-year strategic plan. However, the specific time frames for the activities that would support those strategies are unclear because FDA has not developed a plan that includes actions, priorities, and milestones to implement the strategic plan. The strategic plan states that the FVM Program will develop such an implementation plan, and FDA officials told GAO that they expected to complete one, but as of January 2018, they had not done so. Until the program completes such an implementation plan, it will be difficult for FDA to ensure it is prioritizing and sequencing the necessary actions to achieve the program's objectives.

January 31, 2018

The Honorable Lamar Alexander
Chairman
Committee on Health, Education, Labor, and Pensions
United States Senate

The Honorable Pat Roberts
United States Senate

In January 2011, the FDA Food Safety Modernization Act (FSMA) was signed into law,[1] representing the largest expansion and overhaul of U.S. food safety authorities since the 1930s, according to Department of Health and Human Services' Food and Drug Administration (FDA) documents. FSMA aims to improve food safety by shifting FDA's focus toward preventing food contamination, rather than responding to foodborne illnesses after they have occurred, FDA documents state. The U.S. food supply is generally considered safe, but foodborne illness is still a common and costly—yet largely preventable—public health problem. The Centers for Disease Control and Prevention estimates that every year, as a result of foodborne illness, roughly 1 in 6 Americans (48 million people) gets sick, 128,000 are hospitalized, and 3,000 die.[2] According to a May 2015 estimate by the U.S. Department of Agriculture (USDA), the 15 most common foodborne pathogens in the United States together impose an annual economic burden of over $15.5 billion based on illnesses, hospitalizations, and deaths.[3]

FDA is responsible for overseeing the safety of about 80 percent of the nation's food supply,[4] including animal food.[5] FSMA provides FDA with new enforcement authorities designed to achieve higher rates of

---

[1] Pub. L. No. 111-353, 124 Stat. 3885 (2011).

[2] Elaine Scallan, Robert M. Hoekstra, Frederick J. Angulo, et al., "Foodborne Illness Acquired in the United States—Major Pathogens," *Emerging Infectious Diseases*, vol. 17, no. 1 (2011).

[3] Sandra Hoffmann, Bryan Maculloch, and Michael Batz, "Economic Burden of Major Foodborne Illnesses Acquired in the United States" (Washington, D.C.: Economic Research Service, U.S. Department of Agriculture, May 2015). The economic burden is stated in 2013 dollars. Pathogens are disease-causing organisms, including bacteria such as *Salmonella*, that can be transferred from food animals to the human population; some pathogens can cause foodborne illness.

[4] USDA's Food Safety and Inspection Service is responsible for the safety of meat, poultry, processed egg products, and catfish. FDA is responsible for virtually all other food.

[5] Animal food is made for a variety of species, including livestock from which humans obtain food, pet animals, and laboratory animals. We have previously reported that the safety of animal food is important not only for the health of animals but also for the health of humans. For example, contaminated food for livestock can cause harm both to the livestock and to humans that consume the livestock. See GAO, *Food Safety: FDA Coordinating with Stakeholders on New Rules but Challenges Remain and Greater Tribal Consultation Needed*, GAO-16-425 (Washington, D.C.: May 19, 2016).

compliance with prevention- and risk-based food safety standards and to better respond to and contain problems when they do occur—such as through mandatory recalls of food products that will cause serious adverse health consequences or death. According to a 2016 Congressional Research Service report, FSMA also requires FDA to develop more than 50 regulations,[6] guidelines, and studies to improve food safety.[7] For example, the act requires FDA to develop new, prevention-oriented regulations for the production of foods consumed by humans or animals.

FDA, in addition to being responsible for ensuring the safety of most foods, is responsible for ensuring that those foods are properly labeled. According to FDA officials, the agency conducts a range of activities to improve nutritional status in humans and animals and to reduce the prevalence of diet-related risk factors for chronic disease. These activities include ensuring that food labels for most foods under its jurisdiction contain required nutrition information.[8] As part of its nutrition activities, FDA maintains educational information, databases, and disclosures related to food ingredients.

In fiscal year 2016, FDA centers and offices obligated about $1.3 billion for food safety- and nutrition-related activities.[9] According to the agency, FDA will need additional funding increases to maintain momentum toward full implementation of FSMA. More specifically, FDA has stated that additional funding would assist the agency, its state partners, and the food industry in fully implementing FSMA.[10] In the first

---

[6] We use the terms regulations and rules interchangeably in this chapter.

[7] Congressional Research Service, *Implementation of the FDA Food Safety Modernization Act (FSMA, P.L. 111-353)*, R43724 (Washington, D.C.: Dec. 16, 2016).

[8] Nutrition labeling for raw fruits and vegetables, as well as raw fish, is voluntary.

[9] This $1.3 billion includes obligations for some activities not related to food safety or nutrition, such as those related to FDA's oversight of cosmetics. An obligation is a definite commitment that creates a legal liability of the government to make payment for goods and services received or a legal duty that could mature into a legal liability by virtue of actions beyond the control of the federal government. FDA obligates funds, for example, when it awards a grant, signs a contract, purchases a service, or takes other actions that require the government to make payments. Other federal agencies with food safety and nutrition oversight responsibilities, including USDA, also obligate funds to implement these responsibilities.

[10] Food and Drug Administration, "The Context and History of FSMA Funding" (Washington, D.C.: Feb. 22, 2016), https://www.fda.gov/downloads/Food/GuidanceRegulation/FSMA/UCM436591.pdf, accessed November 20, 2017.

years after FSMA's enactment, FDA was focused on developing key regulations, conducting stakeholder outreach, planning for the implementation of associated rules, investing in state partnerships, and increasing inspection oversight for both domestic and foreign-produced food. FDA estimates that it has received a total net increase of approximately $266 million in additional funding for FSMA implementation in fiscal years 2011 through 2017.

Federal oversight of food safety has been on our list of areas at high risk in the federal government since 2007, when we added it because of risks to the economy, public health, and safety; we have reported that the fragmented federal food safety system has caused inconsistent oversight, ineffective coordination, and inefficient use of resources among the 16 agencies that collectively administer it.[11] In addition, we have found shortcomings in FDA's oversight of seafood, dietary supplements, and other matters.[12] We have made numerous recommendations to improve FDA's oversight, and the agency has implemented some of them. We have also previously found that FDA faced performance management challenges,[13] but the agency has made progress in addressing some of those challenges in recent years. For example, FDA implemented our recommendations to more clearly demonstrate the alignment of activities to strategic goals and to expand its capacity to collect and analyze performance information.

In January 2011, the statutory framework for performance management in the federal government, established by the Government

---

[11] Our high-risk program has focused attention on government operations with greater vulnerabilities to fraud, waste, abuse, and mismanagement or that are in need of transformation to address economy, efficiency, or effectiveness challenges. See GAO, *High-Risk Series: An Update*, GAO-07-310 (Washington, D.C.: January 2007), and *High-Risk Series: Progress on Many High-Risk Areas, While Substantial Efforts Needed on Others*, GAO-17-317 (Washington, D.C.: Feb. 15, 2017).

[12] For example, see GAO, *Imported Seafood Safety: FDA and USDA Could Strengthen Efforts to Prevent Unsafe Drug Residues*, GAO-17-443 (Washington, D.C.: Sept. 15, 2017); *Dietary Supplements: FDA May Have Opportunities to Expand Its Use of Reported Health Problems to Oversee Products*, GAO-13-244 (Washington, D.C.: Mar. 18, 2013); and *Imported Food Safety: FDA's Targeting Tool Has Enhanced Screening, but Further Improvements Are Possible*, GAO-16-399 (Washington, D.C.: May 26, 2016).

[13] GAO, *Food and Drug Administration: Opportunities Exist to Better Address Management Challenges*, GAO-10-279 (Washington, D.C.: Feb. 19, 2010).

Performance and Results Act (GPRA), was updated by the GPRA Modernization Act of 2010 (GPRAMA).[14] As part of our previous work, we examined strategic plans, which are required by GPRAMA. Specifically, the act requires agencies to complete strategic plans in which the agencies define their missions, establish results-oriented goals, and identify objectives and strategies needed to achieve those goals. GPRAMA also requires agencies to use performance information to assess their progress toward strategic goals. GPRAMA requirements apply at the departmental or agency level, not to organizational components. We have previously reported, however, that the requirements can serve as leading practices at other organizational levels within federal agencies, such as component agencies, offices, programs, and projects.[15]

You asked us to examine FDA's food safety- and nutrition-related activities and resources since FSMA's enactment in fiscal year 2011. This chapter examines (1) key food safety- and nutrition-related activities that FDA has conducted since FSMA's enactment and how FDA determined its priorities for those activities; (2) the resources, including personnel, that FDA dedicated to food safety- and nutrition-related activities in fiscal years 2011 through 2016; (3) the extent to which FDA set goals for food safety- and nutrition-related activities in fiscal years 2011 through 2017 and is assessing progress toward those goals; and (4) FDA's planned activities related to food safety and nutrition and the time frames for those activities.

For all objectives, we reviewed documents from FDA and other sources, and we interviewed knowledgeable FDA officials. We generally focused on FDA activities related to food for humans and feed for livestock animals (animal feed), and we generally excluded activities

---

[14] Pub. L. No. 111-352, 124 Stat. 3866 (2011). GPRAMA amends provisions of GPRA, Pub. L. No. 103-62, 107 Stat. 285.

[15] See GAO, *Motor Carriers: Better Information Needed to Assess Effectiveness and Efficiency of Safety Interventions*, GAO-17-49 (Washington, D.C.: Oct. 27, 2016) and *Environmental Justice: EPA Needs to Take Additional Actions to Help Ensure Effective Implementation*, GAO-12-77 (Washington, D.C.: Oct 6, 2011).

concerning veterinary medicine or related substances, such as growth hormones or antibiotics.[16]

To describe key food safety- and nutrition-related activities that FDA has conducted since FSMA's enactment, we reviewed FDA documentation describing actions the agency has undertaken since FSMA's enactment in January 2011 through September 2017. These documents include FDA's *Justification of Estimates for Appropriations Committees* (congressional budget justifications) for fiscal years 2012 through 2018, as well as FDA food safety- and nutrition-related regulations and guidance published from January 2011 through September 2017. To compile and verify lists of key regulations and guidance, we analyzed several sources, including (1) the *Unified Agenda of Federal Regulatory and Deregulatory Actions (Unified Agenda)*, (2) the *Federal Register*,[17] and (3) various FDA websites.[18] In addition, we evaluated FDA's written responses to our questions on this topic, and we updated our lists of key food safety- and nutrition-related regulations and guidance, as appropriate. Using FDA's descriptions of food safety and nutrition—for example, those from FDA's strategic plans and interviews with agency officials—we took several steps to categorize the regulations and guidance as being related to food safety, nutrition, or both. (App. I contains a detailed discussion of the methodology used to compile our lists of key regulations and guidance.)

To assess how FDA determined its priorities for the food safety- and nutrition-related activities it conducted since FSMA's enactment, we interviewed agency officials about their processes for determining food

---

[16] We included activities and resources related to animal drug products and antimicrobial resistance to the extent that they are related to food safety and are included as a subset of the Center for Veterinary Medicine's food safety resources. For example, the use of antimicrobial drugs in feed or water of food-producing animals may affect public health and the risk of antimicrobial resistance development.

[17] The *Unified Agenda*, published biannually by the Office of Management and Budget (OMB), is a compilation of information about federal agencies' planned rulemaking actions. To identify FDA's planned food safety- and nutrition-related rulemaking actions since FSMA's enactment, we reviewed each *Unified Agenda* published from January 2011 through August 2017. To verify the actions taken by FDA, we then compared these planned actions with the draft and final food safety- and nutrition-related regulations published in the *Federal Register*.

[18] The FDA websites include "FSMA Rules & Guidance for Industry," "Food Guidance Documents," and "Search for FDA Guidance Documents."

safety- and nutrition-related priorities, including their decisions to issue regulations and guidance. We then reviewed relevant statutes and regulations, strategic plans, and documentation describing FDA's decision-making processes to verify what agency officials told us. We also compared FDA's processes for determining priorities with commitments that FDA has made as part of an agency initiative to promote transparency and with federal standards for internal controls.[19]

To describe the resources, including personnel, that FDA dedicated to food safety-and nutrition-related activities in fiscal years 2011 through 2016, we collected and analyzed FDA's data on its obligations and full-time equivalent (FTE) staff for these activities during these fiscal years.[20] To assess the reliability of FDA's data, we reviewed FDA's methodology for producing the data, interviewed knowledgeable agency officials, and conducted tests to identify obvious errors, outliers, and missing information. We determined that the data were sufficiently reliable for the purposes of our reporting objectives. Appendix II contains a detailed discussion of the methodology used to compile these data.

To determine the extent to which FDA has set goals for food safety- and nutrition-related activities since fiscal year 2011, we reviewed relevant strategic planning documents, such as those for FDA's Foods and Veterinary Medicine (FVM) Program, which is primarily responsible for carrying out FDA's food safety- and nutrition-related activities. To determine how FDA is assessing its progress toward goals for these activities, we collected and analyzed information on performance measures for which FDA established targets in fiscal year 2017. We also interviewed agency officials about their performance management practices, including the FVM Program's efforts to assess progress, and reviewed leading

---

[19] GAO, *Standards for Internal Control in the Federal Government*, GAO-14-704G (Washington, D.C.: September 2014).

[20] FTEs reflect the total number of regular, straight-time hours (i.e., hours that exclude overtime and holiday hours) worked by employees, divided by the number of compensable hours applicable to each fiscal year. One FTE is equivalent to about 2,080 hours of work. Budget authority is authority provided by federal law for an agency to enter into financial obligations that will result in immediate or future outlays involving federal government funds. During our review, the most recent data available on FDA's budget authority, obligations, and FTEs for food safety and nutrition were for fiscal year 2016.

practices in performance management.[21] Appendix III contains a detailed discussion of the methodology used in our analysis.

To describe FDA's planned activities related to food safety and nutrition and to examine the time frames for those activities, we reviewed documents such as FDA's congressional budget justification for fiscal year 2018,[22] the FVM Program's 10-year strategic plan, and written testimonies by FDA officials. We also compared the extent to which FDA established time frames for activities and goals with leading practices in strategic planning.[23] To identify FDA's planned food safety- and nutrition- related regulations, we reviewed the *Unified Agenda* published in August 2017, and to identify the agency's planned guidance, we analyzed draft guidance that FDA had published since January 2011 that agency officials indicated would be finalized in the future. In addition, we reviewed relevant executive orders and interviewed knowledgeable agency officials to describe the extent to which FDA anticipated that the agency's planned priorities for food safety- and nutrition-related activities might change.

We conducted this performance audit from October 2016 to January 2018 in accordance with generally accepted government auditing standards.

Those standards require that we plan and perform the audit to obtain sufficient, appropriate evidence to provide a reasonable basis for our findings and conclusions based on our audit objectives. We believe that the evidence obtained provides a reasonable basis for our findings and conclusions based on our audit objectives.

---

[21] See GAO, *Agency Performance Plans: Examples of Practices That Can Improve Usefulness to Decisionmakers*, GAO/GGD/AIMD-99-69 (Washington, D.C.: Feb. 26, 1999).

[22] *Department of Health and Human Services, Fiscal Year 2018: Food and Drug Administration, Justification of Estimates for Appropriations Committees.*

[23] See, for example, GAO, *Defense Health Care Reform: Actions Needed to Help Ensure Defense Health Agency Maintains Implementation Progress*, GAO-15-759 (Washington, D.C.: Sept. 10, 2015); *Streamlining Government: Key Practices from Select Efficiency Initiatives Should Be Shared Governmentwide*, GAO-11-908 (Washington, D.C.: Sept. 30, 2011); *Biobased Products: Improved USDA Management Would Help Agencies Comply with Farm Bill Purchasing Requirements*, GAO-04-437 (Washington, D.C.: Apr. 7, 2004); and *Results-Oriented Cultures: Implementation Steps to Assist Mergers and Organizational Transformations*, GAO-03-669 (Washington, D.C.: July 2, 2003).

Background

FDA Organizational Structure for Food Safety and Nutrition

This section provides information on FDA's organizational structure for food safety and nutrition, relevant federal statutes, and processes for developing federal regulations and guidance.

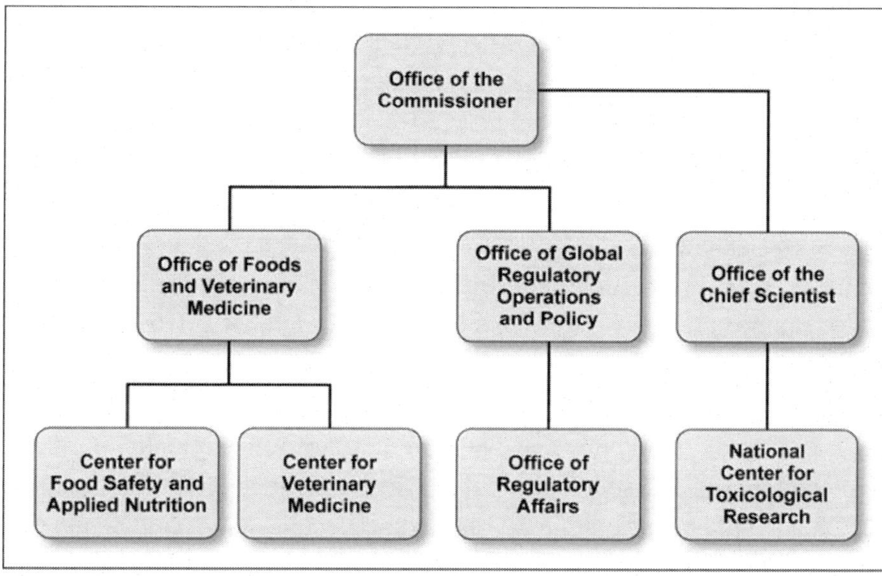

Source: GAO excerpt of Food and Drug Administration (FDA) organizational chart. | GAO-18-174.

Figure 1. FDA Offices and Centers Responsible for Food Safety- and Nutrition-Related Activities.

FDA is responsible for protecting and promoting public health by (1) ensuring the safety of food and animal feed, cosmetics, and radiation-emitting products; (2) ensuring the safety, effectiveness, and security of human and animal drugs, biological products, and medical devices; and (3) regulating tobacco products. Regarding food safety and nutrition specifically, the mission of the agency's FVM Program—overseen by FDA's Office of Foods and Veterinary Medicine—is to promote human health by preventing foodborne illness and fostering good nutrition. The FVM Program contains two components—the Foods Program and the Animal Drugs and Feeds Program—which are managed by the Center for Food Safety and Applied Nutrition and the Center for Veterinary Medicine, respectively. In addition, the Office of Regulatory Affairs is responsible for food safety- and nutrition-related regulatory actions, and the National Center for Toxicological Research conducts research, in part, to improve FDA's ability to assess the safety of regulated products, including food. (See Figure 1 for an excerpt of FDA's organizational chart showing the offices and centers that are responsible for food safety- and nutrition-related activities).

The FDA offices and centers responsible for carrying out food safety- and nutrition-related activities, according to agency documents, are as follows:

- The Office of Foods and Veterinary Medicine is responsible for providing leadership and strategic direction to the FVM Program and for overseeing all Center for Food Safety and Applied Nutrition and Center for Veterinary Medicine activities. In addition, to rapidly detect and respond to major foodborne illness outbreaks, the office is to coordinate activities across FDA field and compliance offices; other federal agencies, such as USDA and the Centers for Disease Control and Prevention; state investigative and laboratory resources; and local resources. The office also is responsible for leading external communications and stakeholder engagement and for coordinating FVM Program-wide resource planning.
- Within the Center for Food Safety and Applied Nutrition, the Foods Program's food safety responsibilities include (1) conducting and supporting food safety research, (2) developing and overseeing enforcement of food safety and quality regulations, (3) coordinating and evaluating FDA's food surveillance and compliance programs, (4) coordinating and evaluating states' food safety activities, and (5) developing and disseminating food safety

and regulatory information to consumers and industry. In addition to improving food safety, the Foods Program carries out nutrition-related activities, such as regulating food labeling and promoting a healthy food supply to reduce the hundreds of thousands of deaths each year attributable to poor diet.

- The Center for Veterinary Medicine administers the Animal Drugs and Feeds Program, whose purposes include (1) protecting the safety of feed and devices for animals and (2) ensuring the safety and effectiveness of animal drugs.
- The Office of Regulatory Affairs carries out field activities—including inspections, sample collections, and import exams—in collaboration with the Center for Food Safety and Applied Nutrition's Foods Program and with the Center for Veterinary Medicine's Animal Drugs and Feeds Program. The Office of Regulatory Affairs is overseen by the Office of Global Regulatory Operations and Policy and is not part of the Office of Foods and Veterinary Medicine.
- The National Center for Toxicological Research conducts peer-reviewed research to advance scientific approaches and tools required to support public health and to improve FDA's ability to assess the safety of regulated products. The center's purpose is to enhance the agency's basis for science-based regulatory decisions by conducting collaborative research to help reduce and rapidly detect contaminants in food products, among other things. The center is overseen by the Office of the Chief Scientist.

Key Food-Related Laws

FDA is responsible for implementing several statutes that have food safety- or nutrition-related elements, including the following:

- Federal Food, Drug, and Cosmetic Act (FFDCA).[24] Under this act and its implementing regulations, FDA oversees the safety of most of the U.S. food supply.[25] Among other things, FDA has the authority under the act to seek an order to remove any food from the market if the food is unsafe. The agency can also pursue enforcement action against those marketing such a food.
- FSMA.[26] This law amended the FFDCA by expanding FDA's food safety authorities and responsibilities in five major areas: (1) requiring the development of prevention-oriented standards for food processing facilities and farms; (2) establishing mandatory inspection frequencies for domestic and foreign food facilities, based on risk; (3) implementing programs to improve oversight of imported products; (4) granting FDA additional enforcement tools, such as mandatory recalls and suspension of registration of food facilities; and (5) mandating that FDA take steps that, when taken, would better integrate its food safety oversight with that of states, localities, tribes, and territories.
- National Nutrition Monitoring and Related Research Act of 1990.[27] Under this act, the Department of Health and Human Services and USDA are required to review and update a report entitled *Dietary Guidelines for Americans*, containing nutritional and dietary information and guidelines for the general public, at least every 5 years. These guidelines, which were most recently issued in December 2015, serve as the basis for federal nutrition policies—such as policies pertaining to updating and expanding food labels. The act requires the guidelines to be based on current scientific and medical knowledge.

---

[24] 21 U.S.C. §§ 301-399g.

[25] USDA, under its authority, is responsible for the safety of meat, poultry, processed egg products, and catfish.

[26] Pub. L. No. 111-353.

[27] Pub. L. No. 101-445, 104 Stat. 1034.

- Nutrition Labeling and Education Act of 1990.[28] This act requires nutrition labeling of most foods and dietary supplements. The act excludes labeling of food sold in restaurants.
- Patient Protection and Affordable Care Act.[29] This 2010 act, among other things, amended the FFDCA in a number of ways, including by mandating that FDA require disclosure of calorie and other nutrition information for standard menu items in certain restaurants and retail food establishments.

Federal Regulation and Guidance Development Processes

Federal regulatory agencies, such as FDA, have authority and responsibility for developing and issuing regulations. Regulations are legally binding and are one of the principal tools that the federal government uses to implement public policy. Agencies' rulemaking processes generally share three basic steps or phases: initiation of rulemaking actions, development of proposed rules, and development of final rules.[30] The Administrative Procedure Act, among other things, establishes the basic process by which all federal agencies develop and issue regulations.[31] The act generally requires agencies to first publish a notice of proposed rulemaking in the *Federal Register*. After giving interested persons an opportunity to comment on the proposed rule, the agency may then publish the final rule.

---

[28] Pub. L. No. 101-535, 104 Stat. 2353.

[29] Pub. L. No. 111-148 § 4205, 124 Stat. 119, 573 (2010).

[30] See GAO, *Federal Rulemaking: Improvements Needed to Monitoring and Evaluation of Rules Development as Well as to the Transparency of OMB Regulatory Reviews*, GAO-09-205 (Washington, D.C.: Apr. 20, 2009).

[31] 5 U.S.C.§ 553. Originally enacted in 1946, the Administrative Procedure Act, ch. 324, 60 Stat. 237 (1946), was repealed and codified in scattered sections of title 5 of the U.S. Code as part of the 1966 revision of title 5. In addition to the requirements under the Administrative Procedure Act, an agency may also need to comply with requirements imposed by other statutes.

Regulatory actions can be categorized as significant—which includes actions that are economically significant—or non-significant under Executive Order 12866.[32] According to the Office of Management and Budget (OMB), the majority of rules issued every year by executive agencies are not significant regulatory actions.

Guidance is not legally binding and is often used by agencies for clarifying or interpreting a statute or regulations.[33] Agencies produce and issue the bulk of guidance—guidance that is considered non-significant—without government-wide standards for those processes. An OMB Bulletin establishes policies and procedures for the development, issuance, and use of "significant" guidance,[34] but most guidance does not fit into this category.[35] Additionally, FDA's Good Guidance Practices regulations differentiate between "Level 1" and "Level 2" guidance and require that, for Level 1 guidance, FDA solicit public comments through a notice in the *Federal Register* and by posting the draft guidance online prior to issuance of the final guidance.

---

[32] Exec. Order No. 12866, *Regulatory Planning and Review* (Washington, D.C.: Sept. 30, 1993).

[33] See GAO, *Regulatory Guidance Processes: Selected Departments Could Strengthen Internal Control and Dissemination Practices*, GAO-15-368 (Washington, D.C.: Apr. 16, 2015).

[34] See Office of Management and Budget, *Final Bulletin for Agency Good Guidance Practices*, 72 Fed. Reg. 3432 (Jan. 25, 2007), and GAO-15-368. The OMB Bulletin directs each agency to develop written procedures for the approval of "significant" guidance, establishes standard elements that must be included in significant guidance, and requires agencies to maintain a website to assist the public in locating significant guidance. Non- significant guidance is not subject to the OMB Bulletin, and guidance procedures are left to agency discretion.

[35] The OMB Bulletin defines "significant" guidance as guidance disseminated to regulated entities or the general public that may reasonably be anticipated to (1) lead to an annual effect on the economy of $100 million or more or adversely affect in a material way the economy, a sector of the economy, productivity, competition, jobs, the environment, public health or safety, or state, local, or tribal governments or communities; (2) create a serious inconsistency or otherwise interfere with an action taken or planned by another agency; (3) materially alter the budgetary impact of entitlements, grants, user fees, or loan programs or the rights and obligations of recipients thereof; or (4) raise novel legal or policy issues arising out of legal mandates, the President's priorities, or the principles set forth in Executive Order 12866, as further amended.

The regulations do not require that FDA solicit public comments for Level 2 guidance.[36] For the purposes of this chapter, "key" guidance includes Level 1 guidance and excludes Level 2 guidance.

During our review, FDA officials noted that regulations and guidance can differ greatly in terms of complexity, detail, and documentation. For example, the officials stated that one final rule took up 258 pages in the *Federal Register* and listed 276 references; as part of this rulemaking, FDA reviewed approximately 300,000 comments.[37] In contrast, the officials stated, another final rule took up fewer than 11 pages in the *Federal Register* and contained 4 references; FDA reviewed 19 comments on this rule.[38]

FDA Has Conducted Numerous Key Food Safety- and Nutrition-Related Activities, Determining Its Priorities Based on Statutes and Strategic Goals

Since FSMA's enactment in January 2011, FDA has conducted numerous key food safety- and nutrition-related activities, determining its priorities for these activities based on statutes and strategic goals. More specifically, FDA issued regulations and guidance,[39] focused mainly on food safety, and conducted other key activities related to food safety and nutrition, such as inspections. FDA prioritized these activities based on statutes—such as FSMA—and its strategic goals, according to agency

---

[36] 21 C.F.R. § 10.115 (2014). These regulations define "Level 1" documents as those that set initial interpretations or more-than-minor changes to interpretations, are scientifically complex, or are likely to be highly controversial; and "Level 2" documents as those that set forth existing practices or minor changes in interpretations. See also GAO-15-368.

[37] Food Labeling: Revision of the Nutrition and Supplement Facts Labels, 81 Fed. Reg. 33,742 (May 27, 2016).

[38] Food Labeling: Nutrient Content Claims; Alpha-Linolenic Acid, Eicosapentaenoic Acid, and Docosahexaenoic Acid Omega-3 Fatty Acids, 79 Fed. Reg. 23,262 (Apr. 28, 2014).

[39] We limited our review to key food safety- and nutrition-related regulations and guidance, which refer to regulations and guidance that are most relevant and important to this review. For the purposes of this chapter, proposed or draft, and subsequent final, versions of the regulations and guidance that were published during the relevant time period were counted only once. See app. I for more information about how we compiled these regulations and guidance documents.

officials, but how the agency decided between issuing regulations and guidance in support of these priorities is unclear because it did not consistently document those decisions.

FDA Published Key Regulations and Guidance, Focused Mainly on Food Safety, and Conducted Other Food Safety- and Nutrition- Related Activities

From January 2011 through September 2017, FDA published key regulations and guidance, focused mainly on food safety, and conducted other food safety- and nutrition-related activities. More specifically, FDA published 33 key proposed or final regulations[40]and 111 key draft or final guidance documents that were related to food safety or nutrition.[41] Of the 33 proposed and final regulations that FDA published, 21 were food safety-related, 7 were nutrition-related, and 5 were related to both.[42] In addition, of the 33 published regulations, 30 regulations were final and 3 were proposed.[43] These 33 regulations included the 7 foundational rules

---

[40] The number of regulations is based on GAO's analysis of the *Unified Agenda*, verified against the *Federal Register* and FDA officials' written and oral responses.

[41] The number of guidance documents is based on GAO's analysis of FDA websites, verified against *Federal Register* publications and FDA officials' written and oral responses. In addition, FDA withdrew certain food safety- or nutrition-related regulations and guidance documents during this time frame. FDA officials told us that the agency withdrew these regulations and guidance documents for various reasons, such as the need for a planned regulation becoming obsolete with the publication of a different rule.

[42] While we could not find an applicable statutory definition for food safety- and nutrition-related activities, FDA officials told us that the agency's food safety activities encompass a range of activities to prevent or mitigate contamination of food for humans and animals.

They noted that FDA's nutrition activities also encompass a range of activities to improve nutritional status in humans and to reduce the prevalence of diet-related risk factors for chronic disease. Some of the agency's activities, however, cut across these two categorizations and, therefore, are related to both food safety and nutrition. FDA officials added that the Center for Veterinary Medicine is not engaged in activities related to the nutritional status of animals and does not receive funding for nutrition activities.

[43] For the purposes of this chapter, "final" includes "interim final" and "final" rules, as indicated in the corresponding *Federal Register* notice (for regulations); for guidance documents, "final" includes documents that are not indicated as being in "draft" form.

required by FSMA.[44] Of the 111 guidance documents that FDA published on its website during the same time frame, 82 were food safety-related, 12 were nutrition-related, and 17 were related to both food safety and nutrition. Furthermore, 80 of those guidance documents were issued in final form, and 31 were published in draft form from January 2011 through September 2017. (See app. I for a list of FDA's key food safety- and nutrition-related regulations and guidance published from January 2011 through September 2017, as well as additional information, including publication dates.)

In addition, among the food safety- and nutrition-related activities that FDA conducted, according to its congressional budget justifications for fiscal years 2012 through 2018, are activities such as conducting compliance inspections and developing risk assessment tools, responding to foodborne illness outbreaks, and providing outreach and education on food safety and nutrition matters.[45] For example, since FSMA's enactment in January 2011, which provided FDA with new enforcement authorities designed to achieve higher rates of compliance with prevention- and risk-based food safety standards, FDA reported conducting risk-based inspections of high-risk and non-high risk food facilities based on the facilities' known safety risks, compliance histories, and years since last inspection.[46]

---

[44] To implement FSMA, FDA published seven critical rules to establish preventive and risk-based standards for the growing, harvesting, packing, processing and distribution of domestic and imported food for people and animals, from farm through transportation to retail. The seven foundational rules are as follows: (1) Preventive Controls for Human Food, (2) Preventive Controls for Animal Food, (3) Produce Safety, (4) Foreign Supplier Verification Program, (5) Accredited Third-Party Certification, (6) Sanitary Transportation, and (7) Intentional Adulteration.

[45] Each year, as part of the annual appropriations process, federal agencies, including FDA, develop and submit to Congress supporting information for their funding requests in congressional budget justifications. FDA's congressional budget justification notes past accomplishments; reflects how FDA proposes to meet its mission, goals, and objectives in the future; and assists Congress in determining how much funding to appropriate for FDA in the upcoming fiscal year.

[46] FDA reported that in fiscal year 2012, it developed a new framework for selecting food facilities for inspection, based on known safety risks, compliance history, and years since the last inspection. Based on these factors, FDA categorized facilities as high-risk and non-high-risk. For example, a facility that manufactures food associated with foodborne outbreaks and class I recalls may be placed in the high-risk category. Likewise, if a facility's inspection results for the previous 5 fiscal years indicate that it has a history of

The agency has also reported developing risk assessment tools—such as FDA-iRISK[47]and the Virtual Deli Risk-Tool[48]—to predict the interventions most effective at preventing foodborne illness. Outbreaks of foodborne illness that the agency reported responding to included outbreaks of *Cyclospora*, *Hepatitis A*, *Listeria monocytogenes*, and *Salmonella*. With respect to outreach and education activities, FDA reported, among other things, conducting consumer studies to learn about attitudes, behaviors, and knowledge of food safety; maintaining a Nutrition Label education program with a television network specializing in cartoons to build awareness of such labels among children; conducting activities to help consumers improve heart health through reduced sodium intake; and partnering with USDA to manage a program for food safety training, education, and outreach to small farm owners and food processors.

Agency Officials Said FDA Based Its Priorities on Statutes and Strategic Goals, but How It Decided between Issuing Regulations and Guidance Was Not Always Clear

Agency Officials Said FDA Based Priorities on Statutes and Strategic Goals

Since FSMA's enactment in January 2011 through fiscal year 2017, FDA based its food safety- and nutrition-related priorities on statutes and

non-compliance with food safety requirements and has food safety violations of regulatory significance, it may be categorized as high-risk. Alternatively, a facility may be categorized as non-high-risk if it neither manufactures a food associated with outbreaks or class I recalls, nor has a compliance history in the previous 5 fiscal years indicating that it has a history of non-compliance with food safety requirements or food safety violations of regulatory significance.

[47] According to FDA's congressional budget justifications for fiscal years 2014 and 2017, in October 2012, FDA released FDA-iRISK—a web-based tool that automates the process of developing mathematical models to simulate risk and intervention in food production chains. This tool also predicts the number of cases of illness prevented by various interventions applied against specific contaminants in specific foods. More recently, FDA released an enhanced version of the tool, with advanced modeling and reporting methods.

[48] According to FDA's congressional budget justification for fiscal year 2016, the Virtual Deli Risk-Tool simulates pathways of *Listeria* cross-contamination in ready-to-eat foods such as those sold in delicatessens.

strategic goals, according to agency officials. However, how the agency decided between issuing regulations and guidance in support of these priorities is unclear because FDA did not consistently document those decisions.[49]

FDA officials told us that the FVM Governance Board, consisting of FDA senior executives, and an executive council,[50] consisting of the deputies for relevant FDA offices and centers, are the internal governing bodies that have been responsible, respectively, for establishing and implementing FVM Program priorities since FSMA's enactment. FDA officials said that the governance board determined top priorities for each fiscal year based, in part, on statutes—including FSMA, FFDCA, and the Patient Protection and Affordable Care Act.[51] (See app. I for the statutory authorities FDA cited in developing the regulations.)

The Bases for FDA's Decisions for Issuing Regulations or Guidance Were Not Always Clear

In determining priorities, FDA officials stated, the board also considered strategic goals, identified in the FVM Program's strategic plan, which the governance board reviewed and approved.[52]

---

[49] Although we examined the period since FSMA's enactment in January 2011 through September 2017, FDA officials told us that they had a decision-making process in place prior to FSMA's enactment to determine food safety- and nutrition-related priorities.

[50] The full name of this executive council is the Food and Veterinary Medicine and the Office of Regulatory Affairs Food and Veterinary Medicine Program Executive Council.

[51] FDA officials informed us that the governance board and executive council were formed in March 2014. The charter for the governance board was signed in March 2014, and the charter for the executive council was signed on various dates in March and April 2014.

[52] To achieve the program's mission over a 10-year time frame, the FVM Program's strategic plan outlines goals, outcomes, objectives, and strategies, taking into account the program's responsibility to implement FSMA and other congressional mandates, and to ensure that FDA regulations and guidance provide clear and reliable direction to industry. See Food and Drug Administration, *FDA Foods and Veterinary Medicine Program Strategic Plan Fiscal Years 2016-2025* (Washington, D.C.: July 14, 2016).

For example, one of the goals in the strategic plan—which provides a framework for implementing FSMA and other legislative authorities—is to protect consumers and animals from foreseeable hazards. A related objective, in support of this goal, is to improve the prevention, detection, and response to foodborne illness outbreaks. Another strategic goal is to foster an environment that promotes healthy and safe food choices, with a related objective to provide and support accurate and useful information and education so consumers can choose healthier diets. FDA officials stated that once the FVM Program top priorities were established for a given fiscal year, the executive council oversaw the implementation of those priorities.

How FDA decided between issuing regulations or guidance in support of food safety- and nutrition-related priorities was not always clear because the agency did not consistently document those decisions. Since January 2011,[53] FDA's decisions to develop either regulations or guidance generally went through a review process whereby FDA considered multiple factors, according to agency officials, on a case-by-case basis.

Specifically, FVM Program officials told us that the decision to develop regulations or guidance was often made by individual offices with the necessary expertise—for example, the Office of Food Safety within the Center for Food Safety and Applied Nutrition or the Office of Surveillance and Compliance within the Center for Veterinary Medicine.[54] This decision was then reviewed by FDA's Office of Policy and, where appropriate, FDA's Office of Chief Counsel. FDA officials told us that among the factors the agency considered were (1) statutory requirements and authorities, (2) the agency's intention to either create a requirement or provide advice, (3) public health and the urgency of a particular

---

[53] As previously noted, we examined the period since FSMA's enactment in January 2011 through September 2017; however, FDA officials told us that the process described here was in place prior to FSMA's enactment.

[54] FDA officials also told us that this decision-making process may differ for broader program initiatives, such as those addressing FSMA and antimicrobial resistance. In those cases, FDA officials told us that cross-cutting groups are formed to develop and review work products. For example, the decision-making and documentation for a specific FSMA regulation would be included in the process of the broader FSMA initiative.

communication, and (4) agency resources.[55] In particular, FDA officials told us that in cases where an underlying statute did not prescribe either issuance of regulations or guidance, the agency developed regulations when it intended for a policy to be legally binding, and it developed guidance when it intended to provide advice to industry or other stakeholders. FDA officials said that more agency resources are needed to develop regulations, compared with guidance; therefore, if the agency believes it can address an issue successfully with guidance, it will do so.

However, FDA's bases for deciding whether to address certain FVM Program priorities through regulations or guidance were not always clear because FDA did not uniformly document its decision-making process—including whether it considered alternative approaches (i.e., developing regulations rather than guidance, or vice versa). According to agency officials, FDA generally had tools in place—such as "concept papers,"[56] and "guidance initiation sheets"[57]—for staff to prepare and present to senior management.[58]

---

[55] See GAO-15-368 for additional information on what we previously found regarding the factors that the Department of Health and Human Services and other agencies (and agency components) considered before deciding whether to issue guidance or undertake rulemaking. Note that FDA was not one of the audited agency components in that GAO report.

[56] For example, the Center for Veterinary Medicine's "Concept or Issue Resolution Paper" template provides instructions on how staff proposing to address an issue through rulemaking should prepare a two- to three-page document describing relevant factors such as why the issue should be addressed through regulation, the agency's legal authority, as well as other options that were considered and why they were rejected.

According to agency officials, the Center for Food Safety and Applied Nutrition also uses concept papers on occasion.

[57] For example, the Center for Veterinary Medicine makes available a "Guidance Information Sheet"—a form consisting of a one-page questionnaire, and one page of instructions. This tool is intended to encourage staff to consider important factors as they develop guidance. This sheet includes instructions for selecting the reason guidance is needed—whether it is to interpret existing regulation or statute, is mandated by a statute, or for another reason.

[58] FDA provided us with blank template forms used for Center for Veterinary Medicine regulations ("CVM Concept or Issue Resolution Paper") and guidance ("CVM Guidance Information Sheet (GIS)") development. The agency also provided us with an example of a completed form, used by the Office of Surveillance and Compliance within the Center for Veterinary Medicine, for developing a guidance related to anti-*Salmonella* food additives in food for animals. However, FDA did not provide us with an example of a completed concept paper, used within the Center for Veterinary Medicine, for developing a food safety- or nutrition-related regulation. Additionally, after multiple requests since June 2017 for documentation of similar forms used within the Center for Food Safety and Applied Nutrition, FDA provided us, in October 2017, with a "CFSAN Concept Paper" template and

These tools were intended, FDA officials told us, to document the reasons for developing a regulation or guidance—and, in some cases, consideration of alternatives—and to ensure agreement with the proposed path forward. However, FDA officials stated that these concept papers and guidance initiation sheets were not uniformly used by Center for Veterinary Medicine or Center for Food Safety and Applied Nutrition staff—and that FDA does not require that the tools be used for every initiated regulation or guidance. These tools were available for staff to use, and FDA officials told us that they were occasionally used for complex decisions and for those they determined to be particularly substantive.

Under its transparency initiative, FDA has committed to improving transparency to regulated industry.[59] In its January 2011 transparency initiative report, FDA describes 19 steps the agency is taking to improve its transparency to regulated industry, including action items and draft proposals for improving transparency by topic area. Among the topic areas discussed is the development of guidance and regulations.

Furthermore, legal scholars and federal courts have noted that it is not always easy to determine whether an agency should address a concern by issuing a regulation—which is subject to public notice and comment requirements under the Administrative Procedure Act—or, instead, by issuing guidance, which is exempt from those requirements.[60]

---

a completed "FSMA Concept Paper" for a FSMA-related interim final rule, used within the Center for Food Safety and Applied Nutrition. During interviews, FVM Program officials provided conflicting responses as to whether concept papers are used within the Center for Veterinary Medicine and the Center for Food Safety and Applied Nutrition.

[59] FDA launched its transparency initiative in response to the President's commitment to openness in government and the Department of Health and Human Services' making transparency a priority. 74 Fed. Reg. 4685 (Jan. 26, 2009), and Department of Health and Human Services, Transparency Task Force, *FDA Transparency Initiative: Improving Transparency to Regulated Industry* (January 2011).

[60] For additional discussion on Administrative Procedure Act requirements, differences between regulations and guidance, and requirements prior to issuance of final guidance, see GAO-15-368. FDA's Good Guidance Practices regulations also stipulate that FDA "may not use documents or other means of communication that are excluded from the definition of guidance document to informally communicate new or different regulatory expectations to a broad public audience for the first time." FDA's Good Guidance Practices also set certain requirements for some guidance documents, including the requirement to solicit public comments through a notice in the *Federal Register* and by posting the draft guidance online prior to issuance of final guidance. 21 C.F.R. § 10.115.

Consequently, agency guidance may be legally challenged based on procedural concerns that the agency inappropriately used guidance, rather than the rulemaking process, or concerns that the agency has issued guidance that goes beyond its authority.[61] For example, agencies have been sued for applying a new interpretation to an existing rule without going through notice and comment rulemaking. In 2015, the U.S. Supreme Court overturned prior federal court rulings that had held that an agency is precluded from substantively changing its interpretation of a rule without notice and comment.[62]

Under federal standards for internal control, management should design control activities to achieve objectives and respond to risks.[63] As an example of designing such control activities, management should clearly document internal control and all transactions and other significant events in a manner that allows the documentation to be readily available for examination.[64] The agency's bases for deciding whether to address certain FVM Program priorities through regulations or guidance would qualify as significant events. Without uniformly documenting the bases for its decisions for issuing either regulations or guidance related to food safety and nutrition—such as by using concept papers or guidance initiation sheets—FDA cannot help ensure consistency and transparency in its decision-making process.

---

[61] See GAO-15-368.

[62] On March 9, 2015, the Supreme Court held in Perez v. Mortgage Bankers Ass'n, No. 13-1041, slip. op (U.S. Mar.9, 2015), that an agency could make substantive changes to an interpretive rule without going through notice and comment under the Administrative Procedure Act. This decision overturned prior federal court rulings that had held that an agency is precluded from substantively changing its interpretation of a regulation through issuance of a new interpretive rule without notice and comment. See, e.g., Alaska Prof'l Hunters Ass'n v. FAA, 177 F.3d 1030 (D.C. Cir. 1999).

[63] GAO-14-704G; see also GAO-15-368.

[64] The use of "significant event" in the context of internal controls differs from the use of "significant" regulations and guidance previously discussed in this chapter.

FDA DEDICATED AT LEAST $1 BILLION PER YEAR, INCLUDING SALARIES FOR AT LEAST 4,300 FTES, TO FOOD SAFETY- AND NUTRITION- RELATED ACTIVITIES IN 2011 THROUGH 2016

For Fiscal Years 2011 through 2016, FDA Dedicated at Least $1 Billion and 4,200 FTEs Annually to Food Safety-Related Activities

FDA has dedicated at least $1.0 billion annually—including salaries for at least 4,300 FTEs—to food safety- and nutrition-related activities in fiscal years 2011 through 2016, with food safety-related activities accounting for about 98 percent of these resources and nutrition-related activities accounting for about 2 percent.[65] For food safety-related activities for this period, FDA dedicated at least $1.0 billion annually, including salaries for at least 4,200 FTEs. In contrast, for nutrition-related activities during this time frame, the agency dedicated at least $20 million annually, including salaries for at least 97 FTEs.

FDA dedicated from $1.0 billion to $1.3 billion annually, including salaries for 4,200 to 4,800 FTEs, to food safety-related activities for fiscal years 2011 through 2016. These resources accounted for about 98 percent of FDA's obligations and FTEs dedicated to food safety- and nutrition-related activities during this time frame.[66]

---

[65] For the purposes of this chapter, "dedicated" is synonymous with "obligated." We excluded user fees from our analysis. Authorized obligations of user fee collections accounted for a small proportion of FDA's total obligations for food safety- and nutrition- related activities in each fiscal year from 2011 through 2016. User fees are assessed to users of goods and services provided by the federal government in conjunction with regulatory activities.

[66] As previously noted, an obligation is a definite commitment that creates a legal liability of the government to make payment for goods and services received or a legal duty that could mature into a legal liability by virtue of actions beyond the control of the federal government. FDA obligates funds, for example, when it awards a grant, signs a contract, purchases a service, or takes other actions that require the government to make payments. During our review, the most recent data available on FDA's obligations and FTEs for food safety and nutrition were for fiscal year 2016.

Table 1. FDA obligations, including salaries for Full-Time Equivalent (FTE) staff, for food safety-related activities, by agency component, for fiscal years 2011 through 2016

Dollars in millions												
	2011		**2012**		**2013**		**2014**		**2015**		**2016**	
Component	**Dollars**	**FTEs**	**Dollars**	**FTEs**	**Dollars**	**FTEs**	**Dollars**	**FTEs**	**Dollars**	**FTEs**	**Dollars**	**FTEs**
Center for Food Safety and Applied Nutrition	242.0	818	252.3	827	234.8	868	253.7	884	266.9	900	285.5	969
Center for Veterinary Medicine	57.9	285	58.4	299	53.0	287	61.5	292	61.5	294	63.5	322
Office of Regulatory Affairs	612.9	2,909	631.3	2,838	577.8	2,870	648.1	2,924	653.7	2,924	733.7	3,054
National Center for Toxicological Research	10.8	48	10.2	46	12.1	56	10.2	47	10.2	45	10.2	49
FDA Headquarters and Infrastructure[a]	154.2	237	159.0	256	162.6	263	181.8	328	187.6	313	196.4	32
Total	**1,077.7**	**4,297**	**1,111.2**	**4,266**	**1,040.3**	**4,344**	**1,155.4**	**4,475**	**1,180.0**	**4,476**	**1,289.0**	**4,718**

Source: GAO analysis of Food and Drug Administration (FDA) data. | GAO-18-174.

Note: Amounts are rounded and may not sum to totals.

[a]FDA Headquarters provides FDA-wide program direction and administrative services. The primary Headquarters components responsible for providing FDA-level oversight and advice are the Office of the Commissioner, the Office of Foods and Veterinary Medicine, the Office of Medical Products and Tobacco, the Office of Operations, and the Office of Global Regulatory Operations and Policy.

Infrastructure refers to FDA's Infrastructure program, which provides office and laboratory space and consists of General Services Administration rental payments; other rent and rent-related activities; and the operations of FDA's White Oak campus, which replaces geographically disparate facilities with new, centralized facilities in Maryland.

Table 1 shows food safety- related obligations and FTEs dedicated by FDA components in each fiscal year (See app. II for more detailed information on FDA components' dedication of resources across specific food safety-related activities).

The resources that these FDA components dedicated to food safety-related activities generally increased or remained the same over this time frame. More specifically:

- Center for Food Safety and Applied Nutrition resources for food safety-related activities generally increased from fiscal years 2011 through 2016. The center dedicated $285.5 million and 969 FTEs in fiscal year 2016 for these activities, compared with $242.0 million and 818 FTEs in fiscal year 2011. Because of data limitations, it is unclear how the center's resources for specific food safety-related activities may have changed from fiscal years 2011 through 2016.[67] However, of the food safety-related activities it carried out in fiscal year 2016, the center obligated the most funding, $111.9 million, to regulatory science activities. According to FDA officials, the center's regulatory science activities included increasing FDA's bioinformatics capacity and developing enhanced screening methods to better detect contaminated food.
- Center for Veterinary Medicine resources for food safety-related activities generally increased from fiscal years 2011 through 2016. The center dedicated $63.5 million and 322 FTEs in fiscal year 2016, compared with $57.9 million and 285 FTEs in fiscal year 2011. FDA reported the center's food safety-related data across two activities: animal food safety and antimicrobial resistance. According to agency officials, the center increased funding for animal food safety activities to implement FSMA. The center's data show a decline in obligations dedicated to antimicrobial

[67] We did not analyze trends related to the Center for Food Safety and Applied Nutrition's obligations and FTEs for specific categories of food safety-related activities because data from fiscal years 2011 through 2013 are not comparable with data from fiscal years 2014 through 2016, as the center made changes to its data reporting systems in fiscal year 2014.

resistance activities from fiscal year 2012 to fiscal year 2013; however, center officials we interviewed stated that this decline was due to a change in the center's methodology for identifying antimicrobial resistance resources.

- Office of Regulatory Affairs resources for food safety-related activities generally increased from fiscal years 2011 through 2016. The office dedicated $733.7 million and 3,054 FTEs to food safety- related activities in fiscal year 2016, compared with $612.9 million and 2,909 FTEs in fiscal year 2011. Over this time frame, the office generally decreased annual resources dedicated to inspections, field examinations, and label reviews. The office increased obligations dedicated to other activities, such as implementing FSMA and conducting FSMA-related training. For example, in July 2017, FDA announced that the office would award $30.9 million to 43 states to implement a new FSMA rule for the safe growing, harvesting, packing, and holding of fruits and vegetables.[68] This followed a separate announcement of office funding to states in 2016. According to an agency website, funding will support states' efforts to formulate multi- year plans to implement a produce safety system and provide related education, outreach, and technical assistance.
- National Center for Toxicological Research resources for food safety-related activities generally remained constant from fiscal years 2011 through 2016. The center dedicated $10.2 million and 49 FTEs to food safety-related activities in fiscal year 2016, compared with $10.8 million and 48 FTEs in fiscal year 2011. FDA reported the center's food safety-related resources across three categories: food, antimicrobial resistance, and dietary supplements. According to FDA officials, the center increased or decreased resources dedicated to specific activities as it initiated or completed various research projects. For example, obligations for dietary supplement research increased from $1.0 million in fiscal

[68] 80 Fed. Reg. 74354 (Nov. 27, 2015) (codified in 21 C.F.R. pt. 112).

year 2015 to $4.0 million in fiscal year 2016 as the center initiated a significant research project on over-the- counter dietary supplements. The center's obligations for other food safety research during this time frame decreased as the center ended multiple, other research projects related to food safety.

- FDA Headquarters and Infrastructure resources for food safety-related activities generally increased from fiscal years 2011 through 2016.This component dedicated $196.4 million and 324 FTEs to food safety-related activities in fiscal year 2016, compared with $154.2 million and 237 FTEs in fiscal year 2011.

In Fiscal Years 2011 through 2016, FDA Dedicated at Least $20 Million and 97 FTEs Annually to Nutrition- Related Activities

FDA dedicated from $20.3 million to $33.7 million annually, including salaries for 97 to 127 FTEs, to nutrition-related activities in fiscal years 2011 through 2016. These resources accounted for about 2 percent of FDA's total obligations and FTEs dedicated to food safety- and nutrition-related activities for this period. Table 2 shows the nutrition-related resources dedicated by each of these components each fiscal year. (See app. II for more detailed information on FDA components' dedication of resources across specific nutrition-related activities).

The resources that these FDA components dedicated to nutrition-related activities generally increased over this time frame, although resources dedicated by some components decreased or remained the same across some fiscal years. More specifically:

- Center for Food Safety and Applied Nutrition resources for nutrition-related activities generally increased from fiscal years 2011 through 2016. The center dedicated $18.5 million and 68 FTEs in fiscal year 2016, compared with $8.2 million and 54 FTEs in fiscal year 2011. Because of data limitations, it is unclear how center resources for specific categories of nutrition-related activities may

- have changed from fiscal years 2011 through 2016.[69] However, of the nutrition-related activities it carried out in fiscal year 2016, the center obligated the most funding, $7.8 million, to regulatory science activities. According to FDA officials, the center's regulatory science activities for nutrition include activities such as studies on consumer perception and evaluation of the intake of nutrients in the diet.
- Office of Regulatory Affairs resources for nutrition-related activities generally decreased from fiscal years 2011 through 2015 but increased from fiscal years 2015 to 2016. The office dedicated $14.8 million and 57 FTEs in fiscal year 2016, compared with $11.8 million and 54 FTEs in fiscal year 2011. FDA officials we interviewed stated that the office increased nutrition-related obligations in fiscal year 2016 to conduct increased field examinations in support of implementing the Food Allergen Labeling and Consumer Protection Act of 2004,[70] which requires food labels to list major allergens present in those foods.[71] In addition, officials stated that the office identified new food risks through these additional field examinations. In response to these risks, the office obligated additional funding in fiscal year 2016 for regulatory actions, such as monitoring class I recalls.[72]

---

[69] We did not analyze trends related to the Center for Food Safety and Applied Nutrition's obligations and FTEs for specific categories of nutrition-related activities because data from fiscal years 2011 through 2013 are not comparable with data from fiscal years 2014 through 2016, as the center made changes to its data reporting systems in fiscal year 2014.

[70] FDA field examinations are on-site examinations of FDA-regulated imported or domestic products. According to FDA officials, the agency may conduct field exams of any commodity or at any location. In contrast, FDA conducts facility inspections ("inspections") at factories, warehouses, and establishments involved in food manufacturing, processing, packing, or holding.

[71] Food Allergen Labeling and Consumer Protection Act of 2004, Pub. L. No. 108-282, tit. II, § 203, 118 Stat. 906, 906-908 (amending 21 U.S.C. 343(w)).

[72] FDA classifies a recall as class I when there is a reasonable probability that the use or exposure to a violative product will cause serious adverse health consequences or death. Examples include food found to contain toxins or undeclared allergens.

Table 2. FDA obligations, including salaries for Full-Time Equivalent (FTE) staff, for nutrition-related activities, by agency component, for fiscal years 2011 through 2016

Dollars in millions												
	2011		**2012**		**2013**		**2014**		**2015**		**2016**	
Component	**Dollars**	**FTEs**	**Dollars**	**FTEs**	**Dollars**	**FTEs**	**Dollars**	**FTEs**	**Dollars**	**FTEs**	**Dollars**	**FTEs**
Center for Food Safety and Applied Nutrition	8.2	54	12.0	55	10.9	55	12.7	59	13.1	61	18.5	68
Office of Regulatory Affairs	11.8	54	11.5	50	9.6	40	10.5	42	10.7	42	14.8	57
FDA Headquarters and Infrastructurea	0.3	2	0.3	2	0.4	2	0.4	3	0.5	3	0.4	2
Total	20.3	110	23.8	107	20.8	97	23.6	104	24.3	106	33.7	127

Source: GAO analysis of Food and Drug Administration (FDA) data. | GAO-18-174.

Note: Amounts are rounded and may not sum to totals.

[a]FDA Headquarters provides FDA-wide program direction and administrative services. The primary Headquarters components responsible for providing FDA-level oversight and advice are the Office of the Commissioner, the Office of Foods and Veterinary Medicine, the Office of Medical Products and Tobacco, the Office of Operations, and the Office of Global Regulatory Operations and Policy.

Infrastructure refers to FDA's Infrastructure program, which provides office and laboratory space and consists of General Services Administration rental payments; other rent and rent-related activities; and the operations of FDA's White Oak campus, which replaces geographically disparate facilities with new, centralized facilities in Maryland.

- FDA Headquarters and Infrastructure resources for nutrition-related activities generally remained the same from fiscal years 2011 through 2016.The component dedicated $0.4 million and 2 FTEs to nutrition- related activities in fiscal year 2016 compared with $0.3 million and 2 FTEs in fiscal year 2011.

FDA Has Set Goals for Food Safety- and Nutrition-Related Activities Every Year since 2011 but Cannot Fully Assess Progress

FDA Has Set Goals for All Agency Activities, and since 2012, the FVM Program Has Set Specific Goals for Food Safety- and Nutrition-Related Activities

Since fiscal year 2011, FDA has set goals for food safety- and nutrition- related activities but has not fully developed the necessary framework to assess progress toward those goals. Specifically, FDA cannot fully assess progress toward its food safety- and nutrition-related goals because the FVM Program has developed performance measures related to some, but not all, of the eight strategic objectives that support its goals. For developed performance measures, the program reports both targets set and measurements taken for specific time frames.

Since fiscal year 2011, FDA has set strategic goals for all agency activities, including food safety- and nutrition-related activities. These goals provide direction to agency programs in their efforts to achieve FDA's public health mission. FDA's 2014-2018 strategic plan identifies four strategic goals: (1) enhance oversight of FDA-regulated products, (2) improve and safeguard access to FDA-regulated products to benefit health, (3) promote better informed decisions about the use of FDA- regulated products, and (4) strengthen organizational excellence and accountability.

These goals encompass FDA's food safety- and nutrition- related activities, in addition to activities not related to food.[73]

Some FDA components have set their own strategic goals for specific product areas they manage in support of FDA's broader goals. For example, the FVM Program set more specific goals for food safety- and nutrition-related activities in fiscal year 2012 and later modified these goals under its current 2016-2025 strategic plan.[74] Under this plan, the FVM Program set two goals for its food safety- and nutrition-related activities: (1) protect American consumers and animals from foreseeable hazards and (2) foster an environment that promotes healthy and safe food choices. The FVM Program identified public health outcomes that it intends to achieve through these goals. It also set several strategic objectives for each goal. The FVM Program's strategic planning framework—including goals, outcomes, and objectives—for food safety- and nutrition-related activities is shown in Table 3. The FVM Program did not have a strategic plan in place in fiscal year 2011, the year of FSMA's enactment; agency officials stated that FDA was preparing to implement FSMA at that time.

According to FVM Program officials, the program's 2016-2025 strategic plan set strategic goals, intended outcomes, and objectives for its food safety- and nutrition-related activities, to better demonstrate the program's work across its core functions, which include food safety and nutrition.[75] Under the older framework used in the 2012-2015 strategic plan, the FVM Program intended to make similar impacts through its food safety- and nutrition-related work, as demonstrated by similarities to the current strategic plan. However, the older framework did not organize the FVM Program's food safety- or nutrition-related activities under one strategic goal specific to either core function.

---

[73] Food and Drug Administration, *FDA Strategic Priorities, 2014-2018* (Washington, D.C.: September 2014).

[74] Food and Drug Administration, *FDA Foods and Veterinary Medicine Program Strategic Plan, Fiscal Years 2016-2025* (Washington, D.C.: July 14, 2016).

[75] According to its 2016-2025 strategic plan, the FVM Program intends to promote public health by preventing foodborne illness, fostering good nutrition, and improving the safety and efficacy of animal health products.

Table 3. Foods and veterinary medicine program strategic planning framework for food safety- and nutrition-related activities, from fiscal years 2016 through 2025 strategic plan

	Food safety	**Nutrition**
Goal	Protect America's consumers and animals from foreseeable hazards	Foster an environment to promote healthy and safe food choices
Outcome	Reduce the incidence of illnesses and deaths attributable to preventable contamination of FDA- regulated food and feed products	Reduce risk factors for and the incidence of nutrition- related chronic disease
Objectives	Establish and gain high rates of compliance with science-based preventive control standards across the global farm-to-table continuum Improve prevention, detection, and response to foodborne illness outbreaks and other food and feed safety incidents Strengthen the ability of consumers to play a proactive role in minimizing food safety risks Enhance the safety of food and feed additives and dietary supplements Strengthen existing partnerships with international, federal, state, local, tribal, and territorial agencies to improve the effectiveness and efficiency of FDA's food safety program for government and industry	Provide and support accurate and useful nutrition information and education, so consumers can choose healthier diets, consistent with the Dietary Guidelines for Americans and other evidence-based recommendations Monitor and assess emerging nutrition science as well as changes in the composition of foods in the marketplace in relation to the nutritional health status of Americans Encourage and facilitate new products and product reformulation to promote a healthier food supply

Source: GAO analysis of the Food and Drug Administration's FDA Foods and Veterinary Program Strategic Plan, Fiscal Years 2016-2025. | GAO-18-174.

FDA Has Developed Performance Measures for Its Food Safety- and Nutrition-Related Goals but Cannot Fully Assess Progress toward These Goals

FDA has developed performance measures for its food safety- and nutrition-related goals, and these measures reflect some, but not all, of the FVM Program's food safety- and nutrition-related objectives supporting these goals. Specifically, FDA has developed performance measures for five of the FVM Program's eight objectives for food safety or nutrition. For example, FDA has developed performance measures for the food safety objective to establish and gain high rates of compliance with science-based preventive control standards; its food safety objective to enhance the safety of food and feed additives and dietary supplements; and its nutrition objective to provide and support accurate and useful nutrition information and education. An example of an FVM Program performance measure that supports the first food safety objective, and which includes targets and time frames, is the measure related to enhancing preventive controls for reducing pathogen contamination of high-risk produce such as tomatoes, cantaloupes, and leafy greens. FDA reports on each measure, either through FDA-TRACK, FDA's agency- wide performance management system, or through its congressional budget justifications.[76]

The FVM Program collects data for completed performance measures to assess progress toward its strategic goals. For developed performance measures, the program reports both targets set and measurements taken for specific time frames. For example, for the FDA-TRACK performance measure "Percentage of health hazard evaluations completed by due date," the program reports 50 percent as its target for the first quarter of fiscal year 2017 and measured 89 percent of health hazard evaluations completed by their due dates. The FVM Program conducts health hazard evaluations to evaluate the scope and severity of hazards posed by foods sourced from animals that may have received unsafe drugs or consumed unsafe feed. By

[76] FDA launched FDA-TRACK in April 2011. According to FDA, performance measures in FDA-TRACK represent the foundational activities and outputs produced by FDA employees.

setting targets and collecting data for this and other performance measures, the FVM Program is assessing progress toward its strategic goals.

FVM Program management reviews developed performance measures through a quarterly performance review process. The quarterly performance review work group includes officials from the Office of Foods and Veterinary Medicine, the Center for Food Safety and Applied Nutrition, and the Center for Veterinary Medicine. According to officials we interviewed, the group reviews all FDA-TRACK and congressional budget justification performance measures related to the FVM Program. The work group examines opportunities to revise these measures in an effort to make progress toward intended public health outcomes, which are reflected in the FVM Program's strategic goals. For example, the performance measure "Average number of working days to serotype priority pathogens in food" supports the FVM Program's strategic goal for food safety.[77] Agency officials we interviewed told us that the work group identified an opportunity to improve the serotyping process in 2010. From fiscal years 2011 through 2017, the work group lowered serotyping targets, collected relevant data, and assessed the program's progress toward improving the serotyping process. Program officials told us that, in addition to the program's quarterly performance review work group, another work group meets twice monthly to review the extent to which FDA performance measures support FSMA requirements and show progress toward FSMA-related public health outcomes.

However, FDA has not developed performance measures for three of its eight food safety- and nutrition-related objectives as of January 2018, and as a result, the FVM Program cannot fully assess progress toward its food safety- and nutrition-related goals. For one of these objectives—the food safety objective related to the consumers' role—FDA developed indicators but did not set targets with time frames.[78] OMB defines a performance measure (which it refers to as a performance indicator) as an

---

[77] According to FDA documentation, faster serotyping enables FDA to more quickly detect foodborne pathogens and effectively respond to foodborne illness outbreaks.

[78] For example, for "the number of Center for Food Safety and Applied Nutrition website page views" indicator, the FVM Program captured quarterly data but did not set a quarterly target for the number of page views it hoped to achieve.

indicator for which a target is set with a time frame, and which will be used to track progress.[79] Therefore, an indicator alone—without targets or time frames—is not a performance measure.[80] For two of the FVM Program's three nutrition objectives—(1) monitor and assess emerging nutrition science as well as changes in the composition of foods in the marketplace in relation to the nutritional health status of Americans, and (2) encourage and facilitate new products and product reformulation to promote a healthier food supply—FDA has not developed an indicator with a target and time frame (See Tables 11 and 12 in app. III for FDA performance measures for food safety- and nutrition-related activities).

FVM Program officials acknowledged the need to develop performance measures that align with those three objectives for which the program has not yet developed measures. Officials told us they expect to complete the development of performance measures for these three food safety- and nutrition-related objectives by the end of the first quarter of fiscal year 2018. Officials we interviewed told us that the program has not yet developed performance measures for nutrition-related activities because management has first focused on updating the food safety-related performance measures since releasing the program's new strategic plan in July 2016. The officials said that management has focused on updating food safety-related performance measures because of the importance of implementing FSMA.

Leading practices in performance management state that federal programs should use performance information to achieve the mission and goals of those programs.[81] As previously noted, the FVM Program's goals are supported by strategic objectives. Strategic objectives serve as the

---

[79] An indicator, according to OMB, is a measurable value that indicates the state or level of something whereas a target is a quantifiable or otherwise measurable characteristic that tells how well or at what level an agency aspires to perform.

[80] Office of Management and Budget, *Preparation, Submission and Execution of the Budget*, OMB Circular No. A-11 (Washington, D.C.: July 2017).

[81] GPRA Modernization Act of 2010 (GPRAMA), Pub. L. No. 111-352, §8, 124 Stat. 3866, 3878 (codified at 31 U.S.C. § 1123(b)(1). Although GPRAMA's requirements apply at the departmental level (e.g., the Department of Health and Human Services), we have previously stated that they can serve as leading practices at other organizational levels, such as component agencies, offices, programs (e.g., the FVM Program), and projects. See, for example, GAO-17-49.

primary unit of analysis for assessment of how a program is achieving its goals,[82] and each objective should be tracked through a set of performance measures that have targets with time frames.[83] Until FDA develops performance measures with associated targets and time frames for all eight of the FVM Program's food safety- and nutrition-related objectives, FDA cannot fully and accurately assess progress toward achieving its food safety- and nutrition-related goals.

FDA's Planned Food Safety- and Nutrition-Related Activities to Focus on FSMA Implementation in 2018, but Time Frames for Longer- Term Activities Are Unclear

FDA's Planned Activities in 2018 Include Enhanced Food Labeling and Continued FSMA Implementation

FDA has identified food safety- and nutrition-related activities that the FVM Program plans to undertake in fiscal year 2018, but its time frames for such activities in the longer term are unclear. According to the *Unified Agenda* of August 2017 and FDA's congressional budget justification for fiscal year 2018, the FVM Program's planned activities for fiscal year 2018 include regulations to enhance food labeling and continued FSMA implementation. However, the specific time frames for activities planned over the longer term are unclear because the FVM Program has not developed an implementation plan that identifies specific actions, priorities, and milestones for its 10-year strategic plan.

FDA's plans for the FVM Program in fiscal year 2018 include publishing regulations to enhance food labeling, continued implementation of FSMA, and other food safety- and nutrition-related activities. For

---

[82] OMB Cir. A-11, p. 210-6.
[83] OMB Cir. A-11, p. 210-7.

example, the *Unified Agenda* published in August 2017[84] identified two food safety- and nutrition-related regulations that FDA planned to publish within 12 months: (1) a proposed labeling rule concerning the use of health claims related to soy protein and coronary heart disease[85] and (2) a final rule on gluten-free labeling of fermented, hydrolyzed, or distilled foods.[86] In addition, the August 2017 *Unified Agenda* identified four food safety- and nutrition-related "inactive rules" that were still being reviewed or considered by FDA at that time.[87]

Furthermore, in September 2017, FDA announced in the *Federal Register* that the agency and many of its components—including the Center for Food Safety and Applied Nutrition and the Center for Veterinary Medicine—were seeking public comments and information from interested parties until December 7, 2017, "to help FDA identify existing regulations and related paperwork requirements that could be modified, repealed, or replaced, consistent with the law, to achieve meaningful burden reduction," while allowing the agency to achieve its

---

[84] A subsequent *Unified Agenda* published in December 2017 provided an update on FDA's planned food safety- and nutrition-related regulations. It is not reflected in our report because of timing constraints.

[85] In October 2017, FDA published a proposed rule to revoke its regulation authorizing the use of health claims on the relationship between soy protein and coronary heart disease on the label or in the labeling of foods. FDA stated that it is taking this action based on its review of the totality of publicly available scientific evidence currently available and its tentative conclusion that such evidence does not support its previous determination that there is significant scientific agreement among qualified experts for a health claim regarding the relationship between soy protein and reduced risk of coronary heart disease. FDA seeks public comments on the proposed rule by Jan. 16, 2018. See 82 Fed. Reg. 50,324 (Oct. 31, 2017).

[86] The August 2017 *Unified Agenda* also identified two "long-term" food safety-related regulations regarding tolerances for residues of new animal drugs in food consumed by humans. The *Unified Agenda* defines long-term actions as items that are under development but for which the agency (in this case, FDA) does not expect to have a regulatory action within 12 months of a given *Unified Agenda* publication.

[87] The August 2017 *Unified Agenda* identified four FDA food safety- and nutrition-related "inactive rules" that were identified in previous editions of the *Unified Agenda* as planned regulations. Those inactive rules were (1) a laboratory accreditation rule, required by FSMA, that would enable FDA to recognize accredited bodies to ensure that food laboratories have appropriate equipment, personnel, and procedures to conduct reliable analyses; (2) a mandate requiring labels on food that has been refused admission into the United States; (3) current good manufacturing practices in the manufacturing, packing, labeling, or holding operations for dietary supplements; and (4) general principles and modernization of food standards.

mission and fulfill statutory obligations. On December 6, 2017, FDA extended the comment period to February 5, 2018. Thus, FDA's past food safety- and nutrition-related regulations were undergoing review as of January 2018.[88]

FDA's planned food safety- and nutrition-related activities also include finalizing draft guidance documents published since FSMA. While considering agency priorities and resources, FDA officials told us they plan to finalize food safety-related guidance such as on (1) mandatory food recalls, (2) current good manufacturing practice requirements for animal food, (3) arsenic in apple juice, and (4) proper labeling of honey and honey products. Regarding nutrition-related guidance, FDA plans to finalize draft guidance such as (1) a list of products for each of the product categories included in the tables of "Reference Amounts Customarily Consumed" relevant to the nutrition facts label, (2) agency views on the scientific evidence needed and the approach to evaluating scientific evidence on the physiological effects of isolated or synthetic non-digestible carbohydrates, and (3) helping vending machine operators and industry better understand and comply with a regulation regarding calorie labeling of food in vending machines.[89]

Other near-term activities for the FVM Program, identified as priorities in FDA's congressional budget justification for fiscal year 2018, include the following activities to be conducted by the Center for Food Safety and Applied Nutrition:

---

[88] For the Center for Food Safety and Applied Nutrition's announcement, see 82 Fed. Reg. 42,503 (Sept. 8, 2017) and for the Center for Veterinary Medicine's announcement, see 82 Fed. Reg. 42,497 (Sept. 8, 2017). For FDA's extension of the comment period to February 5, 2018, see 82 Fed. Reg. 57,560 (Dec. 6, 2017).

[89] In addition, in January 2018, FDA officials informed us that the agency intends to publish guidance in the near future related to FSMA and the nutrition facts label's recently added sugar provision. Regarding FSMA, these guidance documents refer to (1) produce safety implementation and compliance (for non-water provisions), (2) additional chapters of the preventive controls for human foods guidance, (3) preventive controls for animal foods, (4) the Foreign Supplier Verification Program, and (5) "same level of public health protection guidance" related to several FSMA rules. Agency officials also noted that in November 2017, FDA published (1) supplemental, draft guidance on menu labeling for industry and (2) a small entity compliance guide for sanitary transport, related to FSMA.

- working with industry to implement FSMA regulations, such as the produce safety rule;[90]
- responding to foodborne illness outbreaks;
- reviewing infant formula notifications from manufacturers before marketing a new formula;
- helping to ensure the safety of dietary supplements;
- conducting reviews of food ingredients and packaging, such as reviews of food additives and color additives;
- ensuring that foods are safe and properly labeled, including nutrient content and health claims; and
- postmarket monitoring for chemical contaminants.

In addition, FDA's congressional budget justification for fiscal year 2018 states that the Office of Regulatory Affairs will (1) maintain food manufacturing inspections conducted through state contracts and (2) strengthen state, local, and tribal regulatory programs to prevent foodborne illness and to reduce the occurrence of risk factors in retail and food service operations. The fiscal year 2018 document also states that FDA's National Center for Toxicological Research will continue to conduct projects to limit the emergence and spread of drug resistance in bacterial pathogens that compromise the ability to treat foodborne illnesses.

For the Animal Drugs and Feeds Program—another component of the FVM Program, administered by the Center for Veterinary Medicine—FDA's congressional budget justification for fiscal year 2018 states that the agency will continue to conduct, among other things, the following activities:

- advancing FSMA efforts by implementing a modern, prevention-focused, science- and risk-based food and feed safety system;

---

[90] For more information on FDA's Standards for the Growing, Harvesting, Packing, and Holding of Produce for Human Consumption—also known as the produce safety rule or, simply, the produce rule—see GAO, *Food Safety: FDA Continues to Evaluate and Respond to Business Concerns about the Produce Rule*, GAO-18-85 (Washington, D.C.: Nov. 27, 2017) and *Food Safety: FDA's Efforts to Evaluate and Respond to Business Concerns Regarding the Produce Rule*, GAO-17-98R (Washington, D.C.: Nov. 28, 2016).

- surveillance efforts, such as monitoring antimicrobial resistance among intestinal bacteria via a national antimicrobial resistance monitoring system;
- response efforts, such as utilizing laboratory capacity to assist FDA with investigations into potential problems with animal feeds and animal drugs;
- field inspections, investigations, and enforcement activities to ensure the adherence to laws that protect and advance human and animal health;
- enhancing the availability and diversity of safe and effective products that relieve animal pain and suffering, sustain animal health, and do not compromise human health;
- monitoring the safety and effectiveness of animal drugs on the market,
- reviewing "adverse drug experience" reports,
- enforcing compliance actions, and
- reducing the availability of illegally marketed unapproved animal drugs.

Time Frames for Activities Planned for the Longer Term Are Unclear Because the FVM Program's 10- Year Strategic Plan Does Not Have an Implementation Plan

In the longer term, according to FDA officials, the agency plans to pursue the food safety and nutrition strategies that are identified in the FVM Program's 10-year strategic plan.[91] The strategic plan states that FDA places high priority on the implementation of FSMA, which requires sustained focus and commitment over an extended period and focuses on how FDA plans to modernize its food safety work, including:

---

[91] Food and Drug Administration, *FDA Foods and Veterinary Medicine Program Strategic Plan, Fiscal Years 2016-2025* (Washington, D.C.: July 14, 2016).

- an increased focus on obtaining compliance with preventive control standards, rather than finding and responding to violations after an illness or outbreak has occurred;
- strengthening FDA technical expertise and capacity to support FDA and industry in implementing the new prevention standards;
- furthering federal, state, local, and territorial partnerships, and investing in training and capacity to ensure efficient, high-quality, and consistent oversight nationwide; and
- broadening interaction with foreign partners and increasing oversight of importers, who will have more responsibility for the safety of imported foods.

Beyond FSMA, the FVM Program's strategic plan also asserts that it provides greater focus on the agency's public health goals and objectives in the areas of nutrition and chemical safety and that organizational excellence will remain a central strategic priority to make the best use of available resources and to continue investing in FDA's workforce.[92]

The strategic plan also identifies 19 food safety and 9 nutrition strategies that support its food safety- and nutrition-related goals and objectives, as shown in appendix IV. For example, to support the food safety objective of improving prevention, detection, and response to foodborne illness outbreaks and other food and feed safety incidents, the FVM Program has four strategies: (1) improve data analysis and collaboration with food and feed safety partners, including industry, academia, and other domestic and foreign regulatory bodies; (2) improve risk prediction and prioritization to focus FVM Program resources and the efforts of regulatory partners on high-risk commodities and firms; (3) utilize knowledge gained from previous outbreaks and other incidents to better inform risk modeling, identify best practices, and increase compliance through voluntary corrective actions; and (4) improve collaboration and communication within the agency and among external

---

[92] The strategic plan states that FDA conducts a wide range of standard-setting and premarket oversight activities involving chemical safety, encompassing food and feed additives, dietary supplements, food packaging, chemical contaminants in food, and cosmetics.

stakeholders in order to improve the efficiency and effectiveness of FVM Program compliance actions.

To support the nutrition objective of providing and supporting accurate and useful nutrition information and education so that consumers can choose healthier diets consistent with the Dietary Guidelines for Americans and other evidence-based recommendations, the FVM Program has four strategies: (1) improve the availability and accuracy of nutrition information provided to consumers by modernizing nutrition and supplement fact labels, and by expanding nutrition labeling to menus in restaurants, other retail establishments, and vending machines; (2) promote collaboration to advance the science underlying dietary guidance statements, medical food disease claims, and nutrition-related claims; (3) enhance understanding of how consumers notice, understand, and act on labeling and nutrition information, including nutrition facts labels, nutrition product claims, front-of-package nutrition labels, and dietary recommendations; and (4) promote collaboration with stakeholders, including industry, consumer, and public health groups, to enhance consumer nutrition education directed toward age and demographic groups with specific needs.

However, the specific time frames for the activities that would support the strategies identified in its 10-year strategic plan are unclear because the FVM Program has not developed a plan that includes specific actions, priorities, and milestones to implement these strategies. The FVM Program's strategic plan states that an implementation plan will identify the specific actions it will implement over a shorter time frame to achieve the strategic plan's objectives. In addition, the strategic plan states that the implementation plan will prioritize and sequence these actions, based on anticipated resources and other practical constraints, and identify initial and intermediate outcomes and performance measures to monitor progress. FDA officials told us that they expected to complete this implementation plan by the end of fiscal year 2017, but as of January 2018, they had not done so.

Our previous work, consistent with GPRAMA, sets forth several key elements of strategies that can guide agencies in planning and implementing an effective government program.[93] As noted in our prior work, elements such as specific actions, priorities, and milestones are desirable for evaluating progress, achieving results in specific time frames, and ensuring effective oversight and accountability. These elements can also be used to gauge progress when implementing programs and to determine whether adjustments need to be made in order to maintain progress within given time frames.[94] Until FDA completes an implementation plan that includes specific actions, priorities, and milestones for the FVM Program's strategic plan, it will be difficult for the agency to ensure it is prioritizing and sequencing the necessary actions to achieve its objectives.

CONCLUSION

Since FSMA's enactment, FDA has conducted numerous key food safety- and nutrition-related activities. For example, it has issued a number of food safety- and nutrition-related regulations and guidance to implement statutes, such as FSMA, and its strategic goals. FDA considered multiple factors in deciding between developing regulations or guidance, and its decisions went through multiple reviews. However, the agency did not uniformly document the bases for those decisions. Without uniformly documenting the bases for its decisions for issuing either regulations or guidance related to food safety and nutrition—such as by using concept papers or guidance initiation sheets—FDA cannot help ensure consistency and transparency in the decision-making process.

---

[93] GAO, *Electronic Health Records: HHS Strategy to Address Information Exchange Challenges Lacks Specific Prioritized Actions and Milestones*, GAO-14-242 (Washington, D.C.: Mar. 24, 2014), and *Combating Terrorism: Evaluation of Selected Characteristics in National Strategies Related to Terrorism*, GAO-04-408T (Washington, D.C.: Feb. 3, 2004).

[94] See GAO/GGD/AIMD-99-69 and GAO, *Executive Guide: Effectively Implementing the Government Performance and Results Act*, GAO/GGD-96-118 (Washington, D.C.: June 1996).

The FVM Program has set two goals and eight objectives for its food safety- and nutrition-related activities. FDA has developed performance measures to assess progress for some, but not all, of these objectives. Until FDA develops performance measures with associated targets and time frames for all eight of the FVM Program's food safety-and nutrition-related objectives, FDA cannot fully and accurately assess progress toward achieving its food safety- and nutrition-related goals.

The FVM Program's 10-year strategic plan identifies several food safety- and nutrition-related strategies to be pursued by the program, but FDA has not developed an implementation plan—as called for in the strategic plan—that identifies the specific actions it will implement over a shorter time frame to achieve the strategic plan's objectives. Until FDA completes an implementation plan that includes specific actions, priorities, and milestones for the FVM Program's strategic plan, it will be difficult for the agency to ensure it is prioritizing and sequencing the necessary actions to achieve its objectives.

RECOMMENDATIONS FOR EXECUTIVE ACTION

We are making the following three recommendations to FDA:

The Commissioner of FDA should ensure that FVM Program staff uniformly document the bases for their decisions for issuing either regulations or guidance related to food safety and nutrition, such as by using concept papers or guidance initiation sheets. (Recommendation 1)

The Commissioner of FDA should ensure that the FVM Program develops performance measures with associated targets and time frames for all eight of FDA's food safety- and nutrition-related objectives. (Recommendation 2)

The Commissioner of FDA should complete an implementation plan that includes specific actions, priorities, and milestones for the FVM Program's strategic plan. (Recommendation 3)

AGENCY COMMENTS

We are providing a draft of this chapter to the Department of Health and Human Services for review and comment. In its comments, reproduced in appendix V, the Department of Health and Human Services generally concurred with our findings and recommendations and identified several actions that FDA intends to take to address our recommendations, as discussed below. FDA also provided technical comments, which we incorporated as appropriate.

In response to our first recommendation, the Department of Health and Human Services indicated that FDA has established internal documentation and review and clearance procedures for policy documents, including regulations and guidance, to ensure policy documents conform to Administrative Procedure Act requirements, FDA's Good Guidance Practices regulations, and executive orders. The department also indicated that FDA will share templates, such as concept papers, as we recommended, and promote more consistent application across the FVM Program.

In response to our second recommendation, the department indicated that FDA is expanding the use of outcome-based performance measures to monitor and report programmatic impact, such as in implementing FSMA. In addition, the department stated that FDA is using its established processes for developing and refining performance measures to address our recommendation, with the goal of establishing new or refined measures to monitor the outcomes for each goal and objective in the FVM Program's strategic plan. Furthermore, the department stated that, as these measures are operationalized and mature, they will also include targets and timeframes, consistent with our recommendation.

In response to our third recommendation, the department indicated that FDA intends to develop and publish a plan to implement the FVM Program's strategic plan, which will identify specific actions, priorities, and milestones, as we recommended. The department also stated that FDA intends to draft this implementation plan by the fourth quarter of fiscal year 2018.

As agreed with your offices, unless you publicly announce the contents of this chapter earlier, we plan no further distribution until 30 days from the report date. At that time, we will send copies of this chapter to the appropriate congressional committees, the Secretary of Health and Human Services, the Commissioner of FDA, and other interested parties.

Steve D. Morris
Director, Natural Resources and Environment

Appendix I: Key FDA Food Safety-and Nutrition-Related Regulations and Guidance Published from January 2011 through September 2017

Tables 4 and 5 provide information on the key food safety- and nutrition- related regulations and guidance, respectively, that the Food and Drug Administration (FDA) published from January 2011 through September 2017.

To identify these key FDA regulations and guidance, we relied on several sources, including (1) the *Unified Agenda* of *Federal Regulatory and Deregulatory Actions (Unified Agenda)*, (2) the *Federal Register*, and (3) various FDA websites. The *Unified Agenda*, published biannually by OMB, is a compilation of information about federal agencies' planned rulemaking actions. Specifically, to identify FDA's planned food safety- and nutrition-related rulemaking actions since the FDA Food Safety Modernization Act's (FSMA) enactment, we reviewed each *Unified Agenda* published from January 2011 through August 2017. To verify the status of these planned actions, we compared them with the proposed and final food safety- and nutrition-related regulations published in the *Federal Register*. To identify the food safety- and nutrition-related guidance that FDA published from January 2011 through September 2017, we reviewed FDA websites, including "FSMA Rules & Guidance for Industry," "Food Guidance Documents," and "Search for FDA Guidance Documents." In addition, we evaluated FDA's written responses to our questions on this topic and updated our lists of key food safety- and nutrition-related regulations and guidance, as appropriate. For the purposes of this chapter, proposed or draft, and subsequent final versions of the regulations and guidance that were published during the relevant time period were counted only once. In describing FDA's food safety- and nutrition-related activities from January 2011 through September 2017, we use the term "key" to refer to regulations and guidance that are most relevant to and substantive for this review. Specifically, we included "Level 1" guidance, which FDA regulations define as those that set initial interpretations or more-than-minor changes to interpretations, are scientifically complex, or are likely to be highly controversial. We excluded "Level 2" guidance, which FDA regulations define as those that set forth existing practices or minor changes in interpretations.

To categorize the regulations and guidance as being related to food safety, nutrition, or both, we took several steps. Specifically, we analyzed FDA written responses providing descriptions of the terms food safety and nutrition, FDA documents describing the agency's food safety and nutrition activities and goals, and the regulations and guidance. Because we found no applicable statutory definition for food safety- or nutrition-related activities, we relied on FDA's descriptions of food safety and nutrition—for example, those from FDA's strategic plans and interviews with agency officials.[95] We categorized the regulations and guidance as being related to food safety or nutrition depending on whether the regulations or guidance advanced FDA's efforts to achieve the food safety public health outcome or the nutrition public health outcome identified in the Foods and Veterinary Medicine (FVM) Program's strategic plan for fiscal years 2016 through 2025.

---

[95] FDA officials told us that the agency's food safety activities encompass a range of activities to prevent or mitigate contamination of food for humans and animals. They noted that FDA's nutrition activities also encompass a range of activities to improve nutritional status in humans and animals and to reduce the prevalence of diet-related risk factors for chronic disease. Some of the agency's activities, however, cut across these two categories and, therefore, are related to both food safety and nutrition. FDA officials added that the Center for Veterinary Medicine is not engaged in activities related to the nutritional status of animals and does not receive funding for nutrition activities.

Table 4. Key food safety- and nutrition-related regulations published by FDA from January 2011 through September 2017

Title	Subject	Federal Register number	Law(s) authorizing or requiring regulation	Status[a]	Date published
Temperature-Indicating Devices; Thermally Processed Low-Acid Foods Packaged in Hermetically Sealed Containers	FS	76 Fed. Reg. 11,892	Federal Food, Drug, and Cosmetic Act (FFDCA)	Final rule	3/3/2011
Beverages: Bottled Water Quality Standard; Establishing an Allowable Level for Di(2-ethylhexyl)phthalate	FS	76 Fed. Reg.64,810	FFDCA	Final rule	10/19/2011
Animal Food Labeling; Declaration of Certifiable Color Additives	FS	76 Fed. Reg. 71,248	Nutrition Labeling and Education Act (amending FFDCA)	Final rule	11/17/2011
Import Tolerances for Residues of Unapproved New Animal Drugs in Food	FS	77 Fed. Reg. 3653	FFDCA	Proposed rule	1/25/2012
Animal Drugs, Feeds, and Related Products; Regulation of Carcinogenic Compounds in Food-Producing Animals	FS	77 Fed. Reg. 50,591	FFDCA	Final rule	8/22/2012
New Animal Drugs; Updating Tolerances for Residues of New Animal Drugs in Food	FS	77 Fed. Reg. 72,254	FFDCA	Proposed rule	12/5/2012
Criteria Used to Order Administrative Detention of Food for Human or Animal Consumption	FS & N	78 Fed. Reg. 7994	FDA Food Safety Modernization Act (FSMA) (amending FFDCA)	Final rule	2/5/2013
Food Labeling; Gluten-Free Labeling of Foods	FS & N	78 Fed. Reg. 47,154	Food Allergen Labeling and Consumer Protection Act FFDCA	Final rule	8/5/2013
Information Required in Prior Notice of Imported Food	FS	78 Fed. Reg. 32,359	FSMA (amending FFDCA)	Final rule	5/30/2013

Title	Subject	Federal Register number	Law(s) authorizing or requiring regulation	Status[a]	Date published
Establishment, Maintenance, and Availability of Records: Amendment to Record Availability Requirements	FS	79 Fed. Reg. 18,799	FSMA (amending FFDCA)	Final rule	4/4/2014
Food Labeling: Nutrient Content Claims; Alpha-Linolenic Acid, Eicosapentaenoic Acid, and Docosahexaenoic Acid Omega-3 Fatty Acids	N	79 Fed. Reg. 23,262	FFDCA	Final rule	4/28/2014
Current Good Manufacturing Practices, Quality Control Procedures, Quality Factors, Notification Requirements, and Records and Reports, for Infant Formula	FS & N	79 Fed. Reg. 33,057	Infant Formula Act (amending FFDCA)	Final rule	6/10/2014
Food Labeling; Calorie Labeling of Articles of Food in Vending Machines	N	79 Fed Reg. 71,259	Patient Protection and Affordable Care Act (amending FFDCA)	Final rule	12/1/2014
Food Labeling; Nutrition Labeling of Standard Menu Items in Restaurants and Similar Retail Food Establishments	N	79 Fed. Reg. 71,156	Patient Protection and Affordable Care Act (amending FFDCA)	Final rule	12/1/2014
Veterinary Feed Directive	FS	80 Fed. Reg. 31,708	FFDCA	Final rule	6/3/2015
Infant Formula: The Addition of Minimum and Maximum Levels of Selenium to Infant Formula and Related Labeling Requirements	FS & N	80 Fed. Reg. 35,834	FFDCA	Final rule	6/23/2015
Current Good Manufacturing Practice, Hazard Analysis, and Risk-Based Preventive Controls for Food for Animals	FS	80 Fed. Reg. 56,170	FSMA (amending FFDCA) Public Health Service Act	Final rule	9/17/2015

Table 4. (Continued)

Title	Subject	Federal Register number	Law(s) authorizing or requiring regulation	Status[a]	Date published
Current Good Manufacturing Practice, Hazard Analysis, and Risk-Based Preventive Controls for Human Food	FS	80 Fed. Reg. 55,908	FSMA (amending FFDCA) Public Health Service Act	Final rule	9/17/2015
Food Labeling; Gluten-Free Labeling of Fermented or Hydrolyzed Foods	FS & N	80 Fed. Reg. 71,990	Food Allergen Labeling and Consumer Protection Act FFDCA	Proposed rule	11/18/2015
Artificially Sweetened Fruit Jelly and Artificially Sweetened Fruit Preserves and Jams; Revocation of Standards of Identity	N	80 Fed. Reg. 72,581	FFDCA	Final rule	11/20/2015
Accreditation of Third-Party Certification Bodies to Conduct Food Safety Audits and to Issue Certifications	FS	80 Fed. Reg. 74,570	FSMA (amending FFDCA)	Final rule	11/27/2015
Foreign Supplier Verification Programs for Importers of Food for Humans and Animals	FS	80 Fed. Reg. 74,226	FSMA (amending FFDCA) Public Health Service Act	Final rule	11/27/2015
Standards for the Growing, Harvesting, Packing, and Holding of Produce for Human Consumption	FS	80 Fed. Reg. 74,354	FSMA (amending FFDCA) Public Health Service Act	Final rule	11/27/2015
Use of Materials Derived from Cattle in Human Food and Cosmetics	FS	81 Fed. Reg. 14,718	FFDCA	Final rule	3/18/2016
Sanitary Transportation of Human and Animal Food	FS	81 Fed. Reg. 20,092	Sanitary Food Transportation Act FSMA	Final rule	4/6/2016
Antimicrobial Animal Drug Sales and Distribution Reporting	FS	81 Fed. Reg. 29,129	FFDCA	Final rule	5/11/2016
Food Labeling: Revision of the Nutrition and Supplement Facts Labels	N	81 Fed. Reg. 33,742	FFDCA Nutrition Labeling and Education Act	Final rule	5/27/2016

Title	Subject	Federal Register number	Law(s) authorizing or requiring regulation	Status[a]	Date published
Food Labeling: Serving Sizes of Foods That Can Reasonably Be Consumed at One Eating Occasion; Dual-Column Labeling; Updating, Modifying, and Establishing Certain Reference Amounts Customarily Consumed; Serving Size for Breath Mints; and Technical Amendments	N	81 Fed. Reg. 34,000	FFDCA Nutrition Labeling and Education Act	Final rule	5/27/2016
Mitigation Strategies to Protect Food against Intentional Adulteration	FS	81 Fed. Reg. 34,166	FSMA (amending FFDCA)	Final rule	5/27/2016
Amendments to Registration of Food Facilities	FS	81 Fed. Reg. 45,912	FSMA (amending FFDCA) Bioterrorism Act	Final rule	7/14/2016
Substances Generally Recognized as Safe	FS	81 Fed Reg. 54,960	FFDCA	Final rule	8/17/2016
Amendments to Accreditation of Third-Party Certification Bodies to Conduct Food Safety Audits and to Issue Certifications to Provide for the User Fee Program	FS	81 Fed. Reg. 90,186	FSMA (amending FFDCA)	Final rule	12/14/2016
Food Labeling: Health Claims; Dietary Saturated Fat and Cholesterol and Risk of Coronary Heart Disease	N	81 Fed. Reg. 91,716	FFDCA	Interim final rule	12/19/2016

Legend: FS = food safety; N = nutrition; FS & N = food safety and nutrition.

Source: GAO analysis of the Federal Register, the Unified Agenda of Federal Regulatory and Deregulatory Actions, and Food and Drug Administration (FDA) documents. | GAO-18-174.

Note: Regulations establish legally binding requirements and are rooted in agencies' statutory authority. Regulations are included in this table only if they were proposed (with a proposed rule published in the Federal Register) or final (with a final rule or interim final rule published in the Federal Register) during the relevant time frame.

[a]Status indicates the agency "action" published in the Federal Register—final rule, interim final rule, or proposed rule.

Table 5. Key food safety- and nutrition-related guidance published by FDA from January 2011 through September 2017

Title	Subject	Status[a]	Date published
Guidance for Industry: Letter to Firms that Grow, Harvest, Sort, Pack, or Ship Fresh Cilantro	FS	Final	March 2011
Guidance for Industry: Questions and Answers About the Petition Process	FS	Final	April 2011
Guidance for Industry on Fish and Fishery Products Hazards and Controls, Fourth Edition	FS	Final	April 2011
Guidance for Industry: Enforcement Policy Concerning Certain Prior Notice Requirements	FS	Final	June 2011
Food Safety Modernization Act Domestic and Foreign Facility Reinspections, Recall, and Importer Reinspection User Fee Rates for Fiscal Year 2012	FS	Final	August 2011
What You Need to Know about Prior Notice of Imported Food Shipments	FS	Final	August 2011
Guidance for Industry: Measures to Address the Risk for Contamination by Salmonella Species in Food Containing a Pistachio-Derived Product as an Ingredient	FS	Final	September 2011
Guidance for Industry on Studies to Evaluate the Metabolism and Residue Kinetics of Veterinary Drugs in Food-Producing Animals: Comparative Metabolism Studies in Laboratory Animals	FS	Final	September 2011
Guidance for Industry on Studies to Evaluate the Metabolism and Residue Kinetics of Veterinary Drugs in Food-Producing Animals: Metabolism Study to Determine the Quantity and Identify the Nature of Residues	FS	Final	September 2011
Guidance for Industry on Implementation of the Fee Provisions of the FDA Food Safety Modernization Act	FS	Final	October 2011
Guidance for Industry on Evaluating the Safety of Flood-Affected Food Crops for Human Consumption	FS	Final	October 2011
Guidance for Industry: Letter to Firms that Grow, Harvest, Sort, Pack, Process, or Ship Fresh Cantaloupe	FS	Final	November 2011
Guidance for Industry: Prevention of Salmonella enteritidis in Shell Eggs During Production, Storage, and Transportation	FS	Final	December 2011
Guidance for Industry: Questions and Answers Regarding Establishment and Maintenance of Records by Persons Who Manufacture, Process, Pack, Transport, Distribute, Receive, Hold, or Import Food (Edition 5)	FS	Final	February 2012
Guidance for Industry: Testing for Salmonella Species in Human Foods and Direct-Human-Contact Animal Foods	FS	Final	March 2012
Guidance for Industry on the Judicious Use of Medically Important Antimicrobial Drugs in Food-Producing Animals	FS	Final	April 2012
Guidance for Industry: The Seafood List – FDA's Guide to Acceptable Market Names for Seafood Sold in Interstate Commerce	FS & N	Final	July 2012

Title	Subject	Status[a]	Date published
Guidance for Industry: Questions and Answers Regarding the Final Rule, Prevention of Salmonella enteritidis in Shell Eggs During Production, Storage, and Transportation	FS	Final	August 2012
Guidance for Industry: Questions and Answers Regarding Food Facility Registration (Fifth Edition)	FS	Final	December 2012
Guidance for Industry: What You Need to Know About Registration of Food Facilities: Small Entity Compliance Guide	FS	Final	December 2012
Guidance for Industry: A Food Labeling Guide	FS & N	Final	January 2013
Guidance for Industry: What You Need to Know About Administrative Detention of Foods: Small Entity Compliance Guide	FS & N	Final	March 2013
Revised Guidance for Industry on "Studies to Evaluate the Safety of Residues of Veterinary Drugs in Human Food: General Approach to Establish a Microbiological Acceptable Daily Intake (ADI)"	FS	Final	March 2013
Draft Interagency Risk Assessment—Listeria monocytogenes in Retail Delicatessens: Notice of Availability of Documents and Request for Comment	FS	Draft	May 2013
Draft Guidance for Industry on Arsenic in Apple Juice: Action Level; Supporting Document for Action Level for Arsenic in Apple Juice; A Quantitative Assessment of Inorganic Arsenic in Apple Juice	FS	Draft	July 2013
Draft Guidance for Industry: Questions and Answers Regarding the Final Rule, Prevention of Salmonella enteritidis in Shell Eggs During Production, Storage, and Transportation (Layers With Outdoor Access)	FS	Draft	July 2013
Guidance for FDA Staff: Compliance Policy Guide Sec. 690.800 CPG Sec. 690.800 Salmonella in Food for Animals	FS	Final	July 2013
Guidance for Industry: Questions and Answers Regarding Adverse Event Reporting and Recordkeeping for Dietary Supplements as Required by the Dietary Supplement and Nonprescription Drug Consumer Protection Act	FS	Final	September 2013
Draft Risk Profile on Pathogens and Filth in Spices	FS	Draft	November 2013
Guidance for Industry on Purchasing Reef Fish Species Associated with the Hazard of Ciguatera Fish Poisoning	FS	Final	November 2013
Guidance for Industry on New Animal Drugs and New Animal Drug Combination Products Administered in or on Medicated Feed or Drinking Water of Food-Producing Animals: Recommendations for Drug Sponsors for Voluntarily Aligning Product Use Conditions with Guidance for Industry #209	FS	Final	December 2013
Guidance for Industry: Distinguishing Liquid Dietary Supplements from Beverages	FS & N	Final	January 2014
Compliance Policy Guide Regarding Canned Ackee, Frozen Ackee, and Other Ackee Products—Hypoglycin A Toxin	FS	Final	April 2014

Table 5. (Continued)

Title	Subject	Status[a]	Date published
Draft Guidance for Industry: Proper Labeling of Honey and Honey Products	FS & N	Draft	April 2014
Guidance for Industry: Food and Drug Administration Records Access Authority under the Federal Food, Drug, and Cosmetic Act	FS	Final	April 2014
Guidance for Industry: What You Need to Know About Establishment, Maintenance, and Availability of Records: Small Entity Compliance Guide	FS	Final	April 2014
Compliance Policy Guide Regarding Food Facility Registration—Human and Animal Food	FS	Final	June 2014
Environmental Protection Agency and Food and Drug Administration Advice About Eating Fish	FS & N	Draft	June 2014
Guidance for Industry: Considering Whether a Food and Drug Administration-Regulated Product Involves the Application of Nanotechnology	FS & N	Final	June 2014
Guidance for Industry: Assessing the Effects of Significant Manufacturing Process Changes, Including Emerging Technologies, on the Safety and Regulatory Status of Food Ingredients and Food Contact Substances, Including Food Ingredients That Are Color Additives	FS	Final	June 2014
Guidance for Industry: Demonstration of the Quality Factor Requirements for "Eligible" Infant Formulas	N	Final	June 2014
Questions and Answers Regarding Food Facility Registration (Sixth Edition); Guidance for Industry	FS	Final	November 2014
Studies to Evaluate the Metabolism and Residue Kinetics of Veterinary Drugs in Food-Producing Animals: Marker Residue Depletion Studies to Establish Product Withdrawal Periods; Revised Guidance for Industry	FS	Final	March 2015
Studies to Evaluate the Metabolism and Residue Kinetics of Veterinary Drugs in Food-Producing Animals: Validation of Analytical Methods Used in Residue Depletion Studies; Revised Guidance for Industry	FS	Final	March 2015
Guidelines for the Validation of Analytical Methods for the Detection of Microbial Pathogens in Foods and Feeds, 2nd Edition	FS	Final	May 2015
Guidelines for the Validation of Chemical Methods for the FDA FVM Program, 2nd Edition	FS	Final	May 2015
Questions and Answers Regarding Mandatory Food Recalls; Draft Guidance for Industry	FS	Draft	May 2015
Studies to Evaluate the Safety of Residues of Veterinary Drugs in Human Food: Genotoxicity Testing; Revised Guidance for Industry	FS	Final	May 2015
Final Determination Regarding Partially Hydrogenated Oils	FS & N	Final	June 2015

Title	Subject	Status[a]	Date published
Food Allergen Labeling Exemption Petitions and Notifications; Guidance for Industry	FS	Final	June 2015
Guidance for Industry: Cell-Based Products for Animal Use	FS	Final	June 2015
Recommendations for Preparation and Submission of Animal Food Additive Petitions; Guidance for Industry	FS	Final	June 2015
Joint Food and Drug Administration/Health Canada Quantitative Assessment of the Risk of Listeriosis from Soft-Ripened Cheese Consumption in the United States and Canada	FS	Final	August 2015
Use of Nanomaterials in Food for Animals; Guidance for Industry	FS	Final	August 2015
Acceptance Criteria for Confirmation of Identity of Chemical Residues Using Exact Mass Data within the Office of Foods and Veterinary Medicine	FS	Final	September 2015
Draft Compliance Policy Guide: Crotalaria spp. Seeds in Grains	FS	Draft	September 2015
Guidance for Industry: Colored Sea Salt	FS	Final	September 2015
Qualitative Risk Assessment of Risk of Activity/Animal Food Combinations for Activities (Outside the Farm Definition) Conducted in a Facility Co-Located on a Farm	FS	Final	September 2015
Qualitative Risk Assessment of Risk of Activity/Food Combinations for Activities (Outside the Farm Definition) Conducted in a Facility Co- Located on a Farm	FS	Final	September 2015
Veterinary Feed Directive Regulation Questions and Answers; Small Entity Compliance Guide; Guidance for Industry	FS	Final	September 2015
Guidance for Industry: Questions and Answers on FDA's Fortification Policy	N	Final	November 2015
Voluntary Labeling Indicating Whether Food Has or Has Not Been Derived from Genetically Engineered Atlantic Salmon; Draft Guidance for Industry	FS & N	Draft	November 2015
Voluntary Labeling Indicating Whether Foods Have or Have Not Been Derived from Genetically Engineered Plants; Guidance for Industry	FS & N	Final	November 2015
Acrylamide in Foods; Guidance for Industry	FS	Final	March 2016
Ensuring Safety of Animal Feed Maintained and Fed On-Farm; Guidance for Industry	FS	Final	March 2016
Compliance Policy Guide on Crabmeat—Fresh and Frozen— Adulteration with Filth, Involving the Presence of Escherichia coli	FS	Final	April 2016
Exempt Infant Formula Production: Current Good Manufacturing Practices, Quality Control Procedures, Conduct of Audits, and Records and Reports; Guidance for Industry	FS & N	Final	April 2016
Inorganic Arsenic in Rice Cereals for Infants: Action Level; Draft Guidance for Industry	FS	Draft	April 2016

Table 5. (Continued)

Title	Subject	Status[a]	Date published
A Labeling Guide for Restaurants and Retail Establishments Selling Away- From-Home Foods—Part II (Menu Labeling Requirements in Accordance With the Patient Protection Affordable Care Act of 2010); Guidance for Industry	N	Final	May 2016
Frequently Asked Questions About Medical Foods; Second Edition; Guidance for Industry	FS & N	Final	May 2016
Ingredients Declared as Evaporated Cane Juice; Guidance for Industry	N	Final	May 2016
Qualified Facility Attestation Using Form FDA 3942a (for Human Food) or Form FDA 3942b (for Animal Food); Draft Guidance for Industry	FS	Draft	May 2016
Prior Notice of Imported Food Questions and Answers (Edition 3); Guidance for Industry	FS	Final	June 2016
Voluntary Sodium Reduction Goals: Target Mean and Upper Bound Concentrations for Sodium in Commercially Processed, Packaged, and Prepared Foods; Draft Guidance for Industry	N	Draft	June 2016
General Principles for Evaluating the Human Food Safety of New Animal Drugs Used in Food-Producing Animals; Draft Guidance for Industry	FS	Draft	July 2016
The Food and Drug Administration's Policy on Declaring Small Amounts of Nutrients and Dietary Ingredients on Nutrition Labels; Guidance for Industry	N	Final	July 2016
Calorie Labeling of Articles of Food in Vending Machines; Draft Guidance for Industry	N	Draft	August 2016
Classification of Activities as Harvesting, Packing, Holding, or Manufacturing/Processing for Farms and Facilities; Draft Guidance for Industry	FS	Draft	August 2016
Current Good Manufacturing Practice Requirements for Food for Animals; Draft Guidance for Industry	FS	Draft	August 2016
Dietary Supplements: New Dietary Ingredient Notifications and Related Issues; Revised Draft Guidance for Industry	FS	Draft	August 2016
Hazard Analysis and Risk-Based Preventive Controls for Human Food; Draft Guidance for Industry	FS	Draft	August 2016
Human Food By-Products for Use as Animal Food; Draft Guidance for Industry	FS	Draft	August 2016
Compliance Policy Guide Sec. 150.200 Compliance Review of Private Laboratory Analytical Packages (PLAPs)	FS & N	Final	September 2016
Labeling of Infant Formula: Guidance for Industry	FS & N	Final	September 2016
Necessity of the Use of Food Product Categories in Food Facility Registrations and Updates to Food Product Categories (2016 Edition): Guidance for Industry	FS	Final	September 2016
Substantiation for Structure/Function Claims Made in Infant Formula Labels and Labeling: Draft Guidance for Industry	N	Draft	September 2016

Title	Subject	Status[a]	Date published
Use of the Term "Healthy" in the Labeling of Human Food Products: Guidance for Industry	N	Final	September 2016
Veterinary Feed Directive Common Format Questions and Answers; Guidance for Industry	FS	Final	September 2016
Compliance Policy Guide: Section 101.100 FDA Considerations for Recommending Charges under 21 U.S.C.§331(a) or (d) for Causing the Introduction of Violative Products into Interstate Commerce	FS & N	Final	October 2016
Describing a Hazard That Needs Control in Documents Accompanying the Food, as Required by Four Rules Implementing the FDA Food Safety Modernization Act: Guidance for Industry	FS	Draft	October 2016
Frequently Asked Questions About GRAS[b] for Substances Intended for Use in Human or Animal Food: Guidance for Industry	FS	Final	October 2016
Scientific Evaluation of the Evidence on the Beneficial Physiological Effects of Isolated or Synthetic Non-Digestible Carbohydrates Submitted as a Citizen Petition; Draft Guidance for Industry	N	Draft	November 2016
Use of the Seafood List to Determine Acceptable Seafood Names; Draft Compliance Policy Guide	FS & N	Draft	November 2016
Voluntary Qualified Importer Program; Guidance for Industry	FS	Final	November 2016
Compliance Policy Guide: Sec. 615.115 Extralabel Use of Medicated Feeds for Minor Species	FS	Final	December 2016
Preparation of Food Contact Notifications for Food Contact Substances in Contact with Infant Formula and/or Human Milk; Draft Guidance for Industry	FS	Draft	December 2016
Questions and Answers Regarding Food Facility Registration (Seventh Edition); Revised Draft Guidance for Industry	FS	Draft	December 2016
Third-Party Certification Body Accreditation for Food Safety Audits: Model Accreditation Standards; Guidance for Industry and Food and Drug Administration Staff	FS	Final	December 2016
Compliance with and Recommendations for Implementation of the Standards for the Growing, Harvesting, Packing, and Holding of Produce for Human Consumption for Sprout Operations; Draft Guidance for Industry	FS	Draft	January 2017
Control of Listeria monocytogenes in Ready-to-Eat Foods: Revised Draft Guidance for Industry	FS	Draft	January 2017
Draft Guidance for Industry: Study Design Recommendations for Residue Studies in Honey for Establishing Maximum Residue Limits and Withdrawal Periods	FS	Draft	January 2017
Guidance for Industry: Regulation of Intentionally Altered Genomic DNA in Animals: Draft Guidance	FS	Draft	January 2017
Questions and Answers on the Nutrition and Supplement Facts Labels Related to the Compliance Date, Added Sugars, and Declaration of Quantitative Amounts of Vitamins and Minerals; Draft Guidance for Industry	N	Draft	January 2017

Table 5. (Continued)

Title	Subject	Statusa	Date published
Reference Amounts Customarily Consumed: List of Products for Each Product Category; Draft Guidance for Industry	N	Draft	January 2017
Recognition of Acceptable Unique Facility Identifier (UFI) for the Foreign Supplier Verification Programs Regulation: Guidance for Industry	FS	Final	March 2017
Clarification on Food Establishment Waiver from Requirements of the Sanitary Transportation of Human and Animal Food Rule: Guidance for Industry	FS	Final	August 2017
Juice HACCP[c] and the FDA Food Safety Modernization Act: Guidance for Industry	FS	Final	August 2017
Low-Acid Foods Packaged in Hermetically Sealed Containers (LACF) Regulation and the FDA Food Safety Modernization Act: Guidance for Industry	FS	Final	August 2017
Seafood HACCP[c] and the FDA Food Safety Modernization Act: Guidance for Industry	FS	Final	August 2017
Studies to Evaluate the Safety of Residues of Veterinary Drugs in Human Food: General Approach to Establish an Acute Reference Dose	FS	Final	August 2017
Ultrafiltered Milk in the Production of Standardized Cheeses and Related Cheese Products: Guidance for Industry	FS & N	Final	August 2017

Legend: FS = food safety; N = nutrition; FS & N = food safety and nutrition.

Source: GAO analysis of the Federal Register and Food and Drug Administration (FDA) documents. | GAO-18-174.

[a]Status indicates whether the guidance document is in draft or final form. Guidance documents are considered to be final if there is no indication—in the corresponding Federal Register notice or in the guidance document itself—of its being in draft form.

[b]GRAS stands for "generally recognized as safe."

[c]HACCP stands for "Hazard Analysis and Critical Control Point."

These two outcomes were not assumed to be mutually exclusive because some of the agency's activities cut across both the food safety and nutrition categories. Thus, a document we categorized as advancing the food safety public health outcome could also be categorized as advancing the nutrition public health outcome and vice versa. In categorizing the regulations and guidance, we focused on the primary subject matter of the document and did not take into account how resources were dedicated; we acknowledge that resources are dedicated to either food safety or nutrition, but not both. Two analysts independently reviewed and coded the regulations and guidance, and a third analyst conducted a final review in cases where the first and second analysts disagreed. Ultimately, we reached agreement on all categorizations.

The FVM Program's public health outcomes are to be achieved through activities related to food for humans and animals, including food for pets and laboratory animals. However, the scopes of Tables 4 and 5 are limited to the safety and nutrition of food for humans and food-producing animals (livestock such as cattle, poultry, and swine).

Appendix II: FDA Resources for Food Safety- and Nutrition-Related Activities, Fiscal Years 2011 through 2016

Tables 6 through 10 identify the resources—both obligations and full-time equivalents (FTE)—dedicated by three Food and Drug Administration (FDA) centers and one FDA office to specific food safety- and nutrition-related activities in fiscal years 2011 through 2016.

To determine these resources, we obtained and analyzed obligations and FTE data for each relevant center and office that carries out food safety- and nutrition-related activities. Data were obtained directly from FDA databases. Specifically, we requested information on the obligations and FTEs dedicated to various food safety- and nutrition-related activities in each fiscal year from 2011 through 2016—the most recently available data at the time of our report—by four FDA components that conducted food safety- and nutrition-related work: the Center for Food Safety and Applied Nutrition, the Center for Veterinary Medicine, the Office of Regulatory Affairs, and the National Center for Toxicological Research.

These organizational components used time and attendance systems to identify staff hours spent on various food safety- and nutrition-related activities; FDA then converted those hours into FTEs. Specifically:

- The Center for Food Safety and Applied Nutrition used the Resource Reporting System via Project, its voluntary time reporting system, which captures staff hours and resources for various activities.
- The Center for Veterinary Medicine used the Activity Time Reporting system, its mandatory time reporting system.
- The Office of Regulatory Affairs used the Field Accomplishment and Compliance Tracking System—its workflow management system, which captures staff time spent on domestic activities, such as inspections or sample analysis—as well as the Operational and Administrative System for Import Support, which captures staff time on certain import activities, such as sample collection and analysis and field exams.
- The National Center for Toxicological Research used the Project Management System, through which it tracks financial data, and the NCTR Experiment Activity Tracking System, through which it tracks hours worked on various activities.

To determine the obligations for salaries and benefits, FDA officials stated that they compared their FTE data with the agency's Unified Financial Management System. Obligations data include salaries, benefits, and operational funding. We reviewed the methods FDA used to compile the data; interviewed knowledgeable agency officials about the quality and completeness of the estimated data; and conducted tests to identify obvious errors, outliers, and missing information.

Table 6. Obligations, including salaries for Full-Time Equivalent (FTE) staff, dedicated by FDA's center for food safety and applied nutrition to food safety- and nutrition-related activities, fiscal years 2011 through 2013

Dollars in millions							
	Activity	**2011**		**2012**		**2013**	
		Dollars	FTEs	Dollars	FTEs	Dollars	FTEs
Food safety	Detection and compliance	25.4	107	28.3	106	30.2	132
	Prevention and promotion	160.5	518	158.3	521	141.7	488
	Scientific capacity	56.1	193	65.6	200	62.9	248
Food safety total		242.0	818	252.2	827	234.8	868
Nutrition	Detection and compliance	0.4	3	0.5	2	0.4	2
	Prevention and promotion	7.5	49	11.2	51	8.6	45
	Scientific capacity	0.3	2	0.4	2	1.9	8
Nutrition total		8.2	54	12.0	55	10.9	55
Total		250.2	872	264.3	882	245.7	923

Source: GAO analysis of Food and Drug Administration (FDA data). | GAO-18-174.

Note: Amounts are rounded and may not sum totals. In addition, the Center for Food Safety and Applied Nutrition's obligations and FTEs for specific categories of food safety- and nutrition-related activities in fiscal years 2011 through 2013 are not comparable with data from fiscal years 2014 through 2016. These data are not comparable because the center made changes to its time reporting system in fiscal year 2014.

Table 7. Obligations, including salaries for Full-Time Equivalent (FTE) staff, dedicated by FDA's center for food safety and applied nutrition to food safety- and nutrition-related activities, fiscal years 2014 through 2016

Dollars in millions							
	Activity	2014		2015		2016	
		Dollars	FTEs	Dollars	FTEs	Dollars	FTEs
Food safety	Compliance and enforcement	32.5	175	47.2	210	49.6	230
	Education and outreach	19.0	77	22.1	83	34.4	97
	Policy development	15.8	91	20.6	101	21.5	115
	Premarket review	16.1	71	15.2	90	15.0	89
	Regulatory science	96.7	190	108.5	256	111.9	267
	Risk assessment	27.3	116	17.7	51	18.5	54
	Stakeholder engagement	46.3	165	35.6	109	34.6	116
Food safety total		253.7	884	266.9	900	285.5	969
Nutrition	Compliance and enforcement	2.1	8	2.5	14	3.1	18
	Education and outreach	2.2	12	1.2	7	3.3	8
	Policy development	1.5	10	3.2	15	2.8	17
	Premarket review	0.0	0	0.4	0	0.8	0
	Regulatory science	5.2	20	5.2	21	7.8	22
	Risk assessment	0.3	1	0.1	1	0.1	1
	Stakeholder engagement	1.3	7	0.5	3	0.5	3
Nutrition total		12.7	59	13.1	61	18.5	68
Total		266.4	943	280.0	961	304.0	1,037

Source: GAO analysis of Food and Drug Administration (FDA) data. | GAO-18-174.

Note: Amounts are rounded and may not sum totals. In addition, the Center for Food Safety and Applied Nutrition's obligations and FTEs for specific categories of food safety- and nutrition-related activities in fiscal years 2011 through 2013 are not comparable with data from fiscal years 2014 through 2016. These data are not comparable because the center made changes to its time reporting system in fiscal year 2014.

Table 8. Obligations, including salaries for Full-Time Equivalent (FTE) staff, dedicated by FDA's center for veterinary medicine to food safety-related activities, fiscal years 2011 through 2016

Dollars in millions												
	2011		**2012**		**2013**		**2014**		**2015**		2016	
Activity	Dollars	FTEs	Dollars	FTEs	Dollars	FTEs	Dollars	FTEs	Dollars	FTEs	Dollars	FTEs
Animal food safety	37.6	235	38.4	249	36.7	234	45.2	239	45.2	241	47.2	269
Antimicrobial resistance	20.2	50	20.1	50	16.3	53	16.3	53	16.3	53	16.3	53
Total	57.9	285	58.4	299	53.0	287	61.5	292	61.5	294	63.5	322

Source: GAO analysis of Food and Drug Administration (FDA) data. | GAO-18-174.

Note: Amounts are rounded and may not sum to totals.

Table 9. Obligations, including salaries for Full-Time Equivalent (FTE) staff, dedicated by FDA's national center for toxicological research to food safety-related activities, fiscal years 2011 through 2016

Dollars in millions												
	2011		**2012**		**2013**		**2014**		**2015**		**2016**	
Activity	Dollars	FTEs	Dollars	FTEs	Dollars	FTEs	Dollars	FTEs	Dollars	FTEs	Dollars	FTEs
Antimicrobial resistance	3.1	13	1.8	9	2.8	12	2.0	10	2.4	9	2.5	15
Dietary supplements	1.9	7	2.1	10	1.6	3	1.3	2	1.0	2	4.0	10
Other food safety	5.7	28	6.3	27	7.7	41	6.9	35	6.8	34	3.8	24
Total	10.8	48	10.2	46	12.1	56	10.2	47	10.2	45	10.2	49

Source: GAO analysis of Food and Drug Administration (FDA) data. | GAO-18-174.

Note: Amounts are rounded and may not sum to totals.

Table 10. Obligations, including salaries for Full-Time Equivalent (FTE) staff, dedicated by FDA's office of regulatory affairs to food safety- and nutrition-related activities, fiscal years 2011 through 2016

Dollars in millions													
	Activity	2011		2012		2013		2014		2015		2016	
		Dollars	FTEs	Dollars	FTEs	Dollars	FTEs	Dollars	FTEs	Dollars	FTEs	Dollars	FTEs
Food safety	Domestic inspections	163.8	804	153.7	727	117.1	616	116.3	582	120.3	604	140.4	703
	Domestic field exams	35.4	174	32.9	156	33.9	179	34.0	171	31.2	158	34.7	176
	Other domestic activities	133.2	569	155.5	607	148.8	634	201.5	735	216.4	775	256.5	744
	Foreign inspections	12.2	60	16.5	78	15.4	81	16.2	81	15.6	78	14.4	72
	Other foreign activities	11.6	56	12.3	58	14.7	75	16.7	82	16.8	83	24.5	102
	Import field exams	130.4	642	120.2	571	107.3	572	99.6	506	94.7	483	103.0	525
	Import label reviews	27.4	137	22.8	111	18.8	106	18.0	97	16.3	89	18.7	99
	Other import activities, such as laboratory resources	98.9	467	117.3	530	121.9	607	145.8	670	142.4	654	141.6	633
Food safety total		612.9	2,909	631.3	2,838	577.8	2,870	648.1	2,924	653.7	2,924	733.7	3,054
Nutrition	Domestic inspections	1.3	6	1.6	7	1.0	4	1.0	4	1.3	5	2.6	10
	Domestic field exams	0.9	4	0.9	4	0.7	3	0.8	3	1.0	4	2.3	9
	Other domestic activities	—	—	—	—	—	—	.3	1	—	—	.3	1
	Foreign inspections	0.2	1	0.2	1	0.2	1	0.3	1	0.3	1	0.3	1
	Import field exams	2.8	13	2.3	10	2.4	10	2.8	11	2.0	8	4.7	18
	Import label reviews	6.5	30	6.4	28	5.3	22	5.5	22	6.1	24	4.7	18
Nutrition total		11.8	54	11.5	50	9.6	40	10.5	42	10.7	42	14.8	57
Total		624.7	2,963	642.8	2,888	587.4	2,910	658.6	2,966	664.4	2,966	748.5	3,111

Legend: — = not applicable.

Source: GAO analysis of Food and Drug Administration (FDA) data. | GAO-18-174.

Note: Amounts are rounded and may not sum to totals.

When we found errors or missing data, FDA officials corrected these errors and provided us with revised data. Based on FDA's methods for compiling the data, we assessed the reliability of the obligations and FTE data and determined that the data were sufficient and appropriate to support the conclusions reached in this chapter.

Center for Food Safety and Applied Nutrition resources for food safety- related activities include resources for cosmetics-related activities. For budget reporting purposes, the center uses "food safety" to encompass both types of activities. Based on our review of FDA data, we determined that the center's resources for cosmetics-related activities each fiscal year represented a small proportion of the center's reported resources for food safety-related activities.

Center for Veterinary Medicine obligations to food safety-related activities are based on estimates. The center does not track obligations of operational funds by Activity Time Reporting activity codes.

FDA generally considers the FDA Headquarters and Infrastructure's resources to be either for overhead or for supporting other agency activities. FDA Headquarters and Infrastructure data for fiscal years 2014 through 2016 are based on time reporting data, whereas data from fiscal years 2011 through 2013 are based on estimates because of time reporting limitations. We did not analyze FDA Headquarters and Infrastructure resources for specific food safety- and nutrition-related activities because activity-level data were not available from fiscal years 2011 through 2016.

APPENDIX III: FDA PERFORMANCE MEASURESFOR FOOD SAFETY- AND NUTRITION-RELATEDACTIVITIES, FISCAL YEAR2017

This appendix contains information on our analysis of the Foods and Veterinary Medicine (FVM) Program's performance measures in fiscal year 2017.

Table 11. FDA Foods and Veterinary Medicine (FVM) Program Performance Measures for Food Safety-Related Activities in Fiscal Year 2017, by Measure Title and Office or Center

Measure title	Source[a]	FDA office or center
The average number of working days to serotype priority pathogens in food (screening only)	CBJ	Center for Food Safety and Applied Nutrition
Complete review and action on the safety evaluation of direct and indirect food and color additive petitions, within 360 days of receipt	CBJ	Center for Food Safety and Applied Nutrition
GenomeTrackr Program[b]	FDA-TRACK	Center for Food Safety and Applied Nutrition
Human pathogen on plants	FDA-TRACK	Center for Food Safety and Applied Nutrition
New technologies/methodologies to assess chemical food safety	FDA-TRACK	Center for Food Safety and Applied Nutrition
Number of state, local, and tribal regulatory agencies in the U.S. and its territories enrolled in the draft Voluntary National Retail Food Regulatory Program Standards	CBJ	Center for Food Safety and Applied Nutrition
Percentage of "generally recognized as safe" notification reviews completed by Center for Food Safety and Applied Nutrition during the quarter within 180 days of filing	FDA-TRACK	Center for Food Safety and Applied Nutrition
Percentage of analytical package reviews[c] received via FDA's electronic Compliance Management System that meet Office of Compliance timeframes as specified in the quarter	FDA-TRACK	Center for Food Safety and Applied Nutrition
Promote safe production of molluscan shellfish for human consumption: vibrio forecasting and assistance	FDA-TRACK	Center for Food Safety and Applied Nutrition
Reduce the incidence of infection caused by key pathogens commonly transmitted by food: Campylobacter species	CBJ	Center for Food Safety and Applied Nutrition

Table 11. (Continued)

Measure title	Source[a]	FDA office or center
Reduce the incidence of infection caused by key pathogens commonly transmitted by food: Salmonella species	CBJ	Center for Food Safety and Applied Nutrition
Reduce the incidence of infection caused by key pathogens commonly transmitted by food: Shiga toxin-producing Escherichia coli O157:H7	CBJ	Center for Food Safety and Applied Nutrition
Reducing foodborne illness in the population. Work with federal, state, local, tribal, and industry partners, improve preventive controls in food production facilities and reduce the incidence rate (reported cases per 100,000 population per year) of Listeria monocytogenes infections	CBJ	Center for Food Safety and Applied Nutrition
Sequencing-based analytical methods for microbial contaminant detection	FDA-TRACK	Center for Food Safety and Applied Nutrition
Complete review and action on non-administrative abbreviated new animal drug applications and reactivations of such applications received during the fiscal year	CBJ	Center for Veterinary Medicine
Complete review and action on original new animal drug applications and reactivations of such applications received during the fiscal year	CBJ	Center for Veterinary Medicine
Complete review and action on warning letters received within 15 working days to better safeguard our food supply by alerting firms to identified deviations in order to become compliant	CBJ	Center for Veterinary Medicine
Percentage of "generally recognized as safe" notifications reviewed by the Center for Veterinary Medicine within 300 days	FDA-TRACK	Center for Veterinary Medicine
Percentage of health hazard evaluations completed by due date	FDA-TRACK	Center for Veterinary Medicine
Percentage of warning letter recommendation packages reviewed within 25 days	FDA-TRACK	Center for Veterinary Medicine

Measure title	Sourcea	FDA office or center
Total number of collaborating laboratories that will provide coordinated response to high priority chemical and microbial animal feed including pet food contamination events	CBJ	Center for Veterinary Medicine
As required by the FSMA Legislation, cover 100 percent of the High Risk domestic inventory (approximately 19,500 firms) every 3 years	CBJ	Office of Regulatory Affairs
Increase laboratory surge capacity in the event of terrorist attack on the food supply (radiological and chemical samples per week)	CBJ	Office of Regulatory Affairs
Number of domestic and foreign high-risk animal drug and feed inspections	CBJ	Office of Regulatory Affairs
Number of examinations of FDA refused entries[d]	CBJ	Office of Regulatory Affairs
Number of filer evaluations[e]	CBJ	Office of Regulatory Affairs
Number of import food field exams	CBJ	Office of Regulatory Affairs
Number of prior notice import security reviewsf	CBJ	Office of Regulatory Affairs
Number of targeted prohibited bovine spongiform encephalopathy (mad cow disease) inspections	CBJ	Office of Regulatory Affairs

Source: GAO analysis of Food and Drug Administration (FDA) data. | GAO-18-174.

[a]"CBJ" refers to FDA's Justification of Estimates for Appropriations Committees (congressional budget justification) for fiscal year 2017, and FDA-TRACK is FDA's agency-wide performance management system.

[b]Through its GenomeTrackr program, FDA intends to create a network of public health laboratories using whole genome sequencing to investigate foodborne illness outbreaks. Whole genome sequencing provides opportunities to stop outbreaks sooner than possible through traditionally used methods.

[c]FDA uses a network of laboratories to analyze regulated products and monitor compliance with regulations. If laboratory analysis shows a food or dietary supplement is potentially unsafe, FDA reviews the analytical packages to determine if regulatory actions should be taken.

[d]FDA-regulated products can be refused entry in the United States if they do not meet relevant laws and regulations. FDA is responsible for ensuring that these products are either destroyed or exported. In cooperation with U.S. Customs and Border Protection, FDA at times supervises the destruction or exportation of refused entries ("follow-up to refusal").

[e]Import filers send data to FDA, which the agency uses to assess the admissibility of potential imports. FDA conducts filer evaluations on a routine basis to determine whether filers are submitting accurate data. Evaluations are conducted by examining entry documents and comparing the information against data filed by importers.

[f]Importers of human food and animal feed are required to give FDA "prior notice" of their products being offered for entry into the United States. According to FDA, every prior notice is electronically screened. Those identified as high risk receive a security review.

Table 12. FDA Foods and Veterinary Medicine (FVM) program performance measures for nutrition-related activities in fiscal year 2017, by measure title and office or center

Measure title	Source[a]	FDA Office or center
As required by the FDA Food Safety Modernization Act, cover 100 percent of the high-risk domestic inventory (approximately 19,500 firms) every 3 years	CBJ	Office of Regulatory Affairs
Number of examinations of FDA refused entriesb	CBJ	Office of Regulatory Affairs
Number of filer evaluations[c]	CBJ	Office of Regulatory Affairs
Number of import food field exams	CBJ	Office of Regulatory Affairs
Percentage of analytical package reviews[d] received via FDA's electronic Compliance Management System that meet Office of Compliance timeframes as specified in the quarter	FDA-TRACK	Center for Food Safety and Applied Nutrition

Source: GAO analysis of Food and Drug Administration (FDA) data. | GAO-18-174.

[a]"CBJ" refers to FDA's Justification of Estimates for Appropriations Committees (congressional budget justification) for fiscal year 2017, and FDA-TRACK is FDA's agency-wide performance management system.

[b]FDA-regulated products can be refused entry in the United States if they do not meet relevant laws and regulations. FDA is responsible for ensuring that these products are either destroyed or exported. In cooperation with U.S. Customs and Border Protection, FDA at times supervises the destruction or exportation of refused entries ("examinations").

[c]Import filers send data to FDA, which the agency uses to assess the admissibility of potential imports. FDA conducts filer evaluations on a routine basis to determine whether filers are submitting accurate data. Evaluations are conducted by examining entry documents and comparing the information against data filed by importers.

[d]FDA uses a network of laboratories to analyze regulated products and monitor compliance with regulations. If laboratory analysis shows a food or dietary supplement is potentially unsafe, CFSAN reviews the analytical packages to determine if regulatory actions should be taken.

The results of our analysis show how performance measures enable Food and Drug Administration (FDA) officials to assess progress toward the program's strategic goals for food safety- and nutrition-related activities.

We conducted this analysis by evaluating the extent to which each FVM Program performance measure active in fiscal year 2017 aligns with food safety- and nutrition-related strategic objectives identified in the *FDA Foods and Veterinary Medicine Program Strategic Plan, Fiscal Years 2016-2025* (See Tables 11 and 12 for results.)

According to FDA documentation, these objectives, in conjunction with performance measures, provide FVM Program officials with the information they need to assess how the program's efforts are helping it achieve the program's strategic goals.

In identifying the population of performance measures relevant to our analysis, we included only FVM Program performance measures for activities that affect the safety or nutrition of food for humans or food-producing animals.

We excluded from our analysis performance measures for activities that do not affect the safety of foods consumed by humans, such as activities exclusively related to FDA's oversight of pet food.Using a methodology similar to the one described in appendix I, we took several steps to categorize the performance measures as being related to food safety, nutrition, or both. Specifically, two analysts independently coded the performance measures as aligning with one or more of the FVM Program's strategic objectives and intended public health outcomes for food safety and nutrition. A third analyst conducted a final review in cases where the first and second analysts disagreed.

Ultimately, we reached agreement on all categorizations. In Tables 11 and 12, we list the performance measures related to food safety and nutrition, respectively.

Based on our analysis, some performance measures align with both food safety and nutrition objectives. For example, we have determined that the performance measure related to the "number of import food field exams" aligns with the three objectives related to preventive control standards, additives and dietary supplements, and nutrition information and education. We made this determination because field exams likely contribute to higher compliance rates with food safety preventive control standards, include checks for non-permitted additives, and include checks for labeling compliance. Measures that align with both food safety and nutrition objectives are included in both Tables 11 and 12.

APPENDIX IV: FDA FOODS AND VETERINARY MEDICINE PROGRAM'S GOALS, OBJECTIVES, AND STRATEGIES FOR FOOD SAFETY AND NUTRITION, FISCAL YEARS 2016 THROUGH 2025

Table 13 shows the Food and Drug Administration (FDA) Foods and Veterinary Medicine Program's goals, objectives, and strategies for food safety and nutrition, according to the program's strategic plan for fiscal years 2016 through 2025.

Table 13. Food and Drug Administration's (FDA) Foods and Veterinary Medicine (FVM) Program's Goals, Objectives, and Strategies for Food Safety and Nutrition, Fiscal Years 2016 through 2025

Goal	Objective	Strategy
Food safety	Establish and gain high rates of compliance with science-based preventive control standards across the global farm- to-table continuum.	Develop and implement a robust communication, training, and technical assistance plan for preventive control standards.
		Utilize innovative inspection and compliance strategies, backed by administrative and judicial enforcement tools to foster compliance with preventive control standards.
		Evaluate and improve the effectiveness of preventive control standards.
		Evaluate and mitigate the risks of chemical exposures in food and feed products that pose public health or regulatory concerns.
		Evaluate and mitigate the risks of microbiological hazards in food and feed products.
	Improve prevention, detection, and response to foodborne illness outbreaks and other food and feed safety incidents.	Improve data analysis and collaboration with food and feed safety partners, including industry, academia, and other domestic and foreign regulatory bodies.
		Improve risk prediction and prioritization to focus FVM resources and the efforts of regulatory partners on high-risk commodities and firms.
		Utilize knowledge gained from previous outbreaks and other incidents to better inform risk modeling, identify best practices, and increase compliance through voluntary corrective actions.
		Improve collaboration and communication within the agency and among external stakeholders in order to improve the efficiency and effectiveness of FVM compliance actions.

Goal	Objective	Strategy
Food safety	Strengthen the ability of consumers to play a proactive role in minimizing food safety risks.	Increase research, data analysis, and systematic evaluation to improve the effectiveness of food safety education in changing unsafe consumer food-handling behaviors.
		Increase consumer-based communications and outreach regarding safe food-handling practices, including leveraging a variety of community-based education programs.
		Enhance communication to consumers during and following illness outbreaks.
	Enhance the safety of food and feed additives and dietary supplements.	Implement innovative regulatory and compliance strategies to improve pre-market oversight and safety evaluation of food and feed additives, generally regarded as safe substances, and FDA's ability to verify that substances added to the food supply meet applicable safety standards.
		In collaboration with external stakeholders, including regulatory and scientific partners, improve data-driven, post-market surveillance of substances added to the food supply to understand and assess changing use and intake patterns, emerging toxicological data, and adverse event reports.
		Make innovative use of FDA resources and collaborative initiatives with regulatory partners and industry to achieve the goals of FDA's dietary supplement Good Manufacturing Practice regulation and other standards related to the safety, quality, identity, and integrity of dietary supplements.
		Expand consumer and health care provider education regarding safe use of dietary supplements.
	Strengthen existing partnerships with international, federal, state, local, tribal, and territorial agencies to improve the effectiveness and efficiency of the FDA's food safety program for government and industry.	Enhance leveraging of food safety efforts with domestic and international regulatory and public health partners.
		Increase engagement and partnerships with domestic and international intergovernmental organizations to strengthen food safety capacity-building and food safety standards development.

Table 13. (Continued)

Goal	Objective	Strategy
Food safety	Strengthen existing partnerships with international, federal, state, local, tribal, and territorial agencies to improve the effectiveness and efficiency of the FDA's food safety program for government and industry.	Enhance leveraging of food safety efforts with domestic and international regulatory and public health partners.
		Increase engagement and partnerships with domestic and international intergovernmental organizations to strengthen food safety capacity-building and food safety standards development.
		Utilize the Partnership for Food Protection[a] and its fiscal year 2015- 2020 strategic plan to continue to build the national Integrated Food Safety System[b] domestically.
Nutrition	Provide and support accurate and useful nutrition information and education so consumers can choose healthier diets consistent with the Dietary Guidelines for Americans and other evidence-based recommendations.	Improve the availability and accuracy of nutrition information provided to consumers by modernizing nutrition and supplement fact labels, and by implementing the expansion of nutrition labeling to menus in restaurants, other retail establishments, and vending machines.
		Promote collaboration to advance the science underlying dietary guidance statements, medical food disease claims, and nutrition- related claims.
		Enhance understanding of how consumers notice, understand, and act on labeling and nutrition information, including nutrition facts labels, nutrition product claims, front-of-package nutrition labels, and dietary recommendations.
		Promote collaboration with stakeholders, including industry, consumer, and public health groups, to enhance consumer nutrition education directed toward age and demographic groups with specific needs.
		Enhance the FVM Program's capacity to gather and analyze information about the composition and labeling of foods in the marketplace.
		Collaborate with federal partners who conduct food intake surveys to monitor overall intake of nutrients and food-related components in the U.S. population, and assess the effect of FVM initiatives on population intake (overall and in subpopulations of interest), risk factors for chronic disease, and health outcomes.

Goal	Objective	Strategy
Nutrition		Enhance the FVM Program's capacity to review and respond to emerging scientific and technological issues in food and nutrition.
	Encourage and facilitate new products and product reformulation to promote a healthier food supply	Improve the nutritional profile of foods through mechanisms such as labeling, voluntary industry guidelines, and utilizing additional regulatory authorities where appropriate.
		Collaborate with stakeholders to encourage research into healthful ingredient substitutes to support development and reformulation of healthier food options.

Source: FDA Foods and Veterinary Medicine Program Strategic Plan, Fiscal Years 2016-2025. | GAO-18-174.

[a]The Partnership for Food Protection is a collaborative partnership of federal, tribal, state, and local governments that is intended to develop and implement procedures, best practices, and other work products that would advance integration of the U.S. food safety system.

[b]The FDA Food Safety Modernization Act, which was signed into law in January 2011, provided a mandate for FDA to build a national, integrated food safety system by taking steps to better integrate its food safety oversight with that of states, localities, tribes, and territories.

Appendix V: Comments from the Department of Health and Human Services

DEPARTMENT OF HEALTH & HUMAN SERVICES OFFICE OF THE SECRETARY

Assistant Secretary for Legislation
Washington, DC 20201

JAN 1 8 2018

Steve Morris
Director, Natural Resources and Environment
U.S. Government Accountability Office
441 G Street NW
Washington, DC 20548

Dear Mr. Morris:

Attached are comments on the U.S. Government Accountability Office's (GAO) report entitled, "*Food Safety and Nutrition: FDA Can Build on Existing Efforts to Measure Progress and Implement Key Activities*" (GAO-18-174).

The Department appreciates the opportunity to review this report prior to publication.

Sincerely,

Barbara Pisaro Clark

Barbara Pisaro Clark
Acting Assistant Secretary for Legislation

GENERAL COMMENTS OF THE DEPARTMENT OF HEALTH AND HUMAN SERVICES (HHS) ON THE GOVERNMENT ACCOUNTABILITY OFFICE'S DRAFT REPORT ENTITLED - FOOD SAFETY AND NUTRITION: FDA CAN BUILD ON EXISTING EFFORTS TO MEASURE PROGRESS AND IMPLEMENT KEY ACTIVITIES (GAO-18-174)

The U.S. Department of Health and Human Services (HHS) appreciates the opportunity from the Government Accountability Office (GAO) to review and comment on this draft report. The Food and Drug Administration (FDA) is committed to delivering public health outcomes by modernizing the food safety system and empowering consumers to make healthier food choices. FDA is making good progress moving towards a preventive food safety framework with the implementation of the FDA Food Safety Modernization Act (FSMA). In addition, implementation of the new Nutrition Facts label for packaged foods will make it easier for consumers to make better informed food choices. The Agency is also restructuring its Office of Regulatory Affairs, implementing a commodity-based and vertically-integrated structure, to streamline regulatory programs and better deliver public health outcomes.

As FDA continues modernization efforts, the Agency is committed to having clear implementation plans for its strategic goals and objectives and communicating its path forward to the public. FDA is also modernizing its metrics to focus on measures that demonstrate enhanced health outcomes for consumers. FDA plans to use this work to mature from primarily process-focused performance measures to outcome-based performance

Recommendation 1
The Commissioner of FDA should ensure that program staff uniformly document the bases for their decisions for issuing either regulations or guidance related to food safety and nutrition, such as using concept papers or guidance initiation sheets.

HHS Response
HHS concurs with GAO's recommendation.

FDA has established internal documentation and review and clearance procedures for policy documents, including regulations and guidances, to ensure policy documents conform to Administrative Procedure Act requirements, FDA's Good Guidance Practices regulations, and Executive Orders. FDA will continue to work with its offices to share templates, such as concept papers, and promote more consistent application across the Food and Veterinary Medicine (FVM) Program.

Recommendation 2
The Commissioner of FDA should ensure that program develops performance measures with associated targets and time frames for all eight of FDA's food safety-and nutrition-related objectives.

HHS Response
HHS concurs with GAO's recommendation.

FDA uses outcome-based performance measures for the preparation, submission, and execution of the budget. As a long-term commitment to effective management of FDA programs, FDA is expanding the use of outcome-based performance measures to monitor and report programmatic

GENERAL COMMENTS OF THE DEPARTMENT OF HEALTH AND HUMAN SERVICES (HHS) ON THE GOVERNMENT ACCOUNTABILITY OFFICE'S DRAFT REPORT ENTITLED - FOOD SAFETY AND NUTRITION: FDA CAN BUILD ON EXISTING EFFORTS TO MEASURE PROGRESS AND IMPLEMENT KEY ACTIVITIES (GAO-18-174)

impact. For example, FDA developed frameworks to define the intermediate and strategic outcomes that inform planning and illustrate the successful implementation of FSMA. The resulting performance measures monitor progress towards the accomplishment of the FSMA public health outcome objective of the American food supply becoming safer. This expanded use of outcome-based performance measures is a result of the FDA's commitment to the effective management of available resources to provide the maximum public health gains.

In 2016, FDA began the process of refining, developing and maturing performance measures aligned to the FVM Strategic Plan and FSMA. Upon completing the transition to the new FVM Program Strategic Plan for Fiscal Years 2016-2025, FDA's existing performance management processes identified that several of FVM's strategic objectives – both food safety and nutrition-related – did not have performance measures. Also identified was the need to mature from primarily process, or output, performance measures to outcome-based performance measures. By addressing both topics, FDA will increase its ability to track and monitor progress for both the FVM Strategic Plan and FSMA. FDA is using its established processes for the development and refinement of performance measures to address this recommendation with the goal of establishing new or refined measures to monitor the outcomes for each goal and objective in the FVM Strategic Plan. As these measures are operationalized, they will be integrated into existing performance management processes and will, as the measures mature, include targets and timeframes.

Recommendation 3
The Commissioner of FDA should complete an implementation plan that includes specific action, priorities, and milestones for the program's strategic plan.

HHS Response
HHS concurs with GAO's recommendation.

In 2018, FDA is committed to the finalization of a recurring implementation planning process that will be used to develop and publish a report. The implementation plan will identify the actions that the FVM Program will execute to accomplish the objectives identified in the FVM Strategic Plan. It will include a prioritization of specific activities and include milestones to achieve over a shorter time span. Additionally, the implementation plan will identify initial and immediate results that will align with existing FDA strategic planning and performance management processes to develop or refine performance measures to ensure that progress can be monitored for each strategic objective. FDA intends to have a draft implementation plan by the 4th quarter of FY 2018.

In: Food Safety
Editor: Georges Martin

ISBN: 978-1-53616-762-7

Chapter 2

FOOD SAFETY: FEDERAL EFFORTS TO MANAGE THE RISK OF ARSENIC IN RICE*

United States Government Accountability Office

ABBREVIATIONS

ARS	Agricultural Research Service
CDC	Centers for Disease Control and Prevention
EPA	Environmental Protection Agency
FDA	Food and Drug Administration
FGIS	Federal Grain Inspection Service
HHS	Department of Health and Human Services
NIFA	National Institute of Food and Agriculture
NIH	National Institutes of Health
NRC	National Research Council
ppb	parts per billion

* This is an edited, reformatted and augmented version of United States Government Accountability Office, Report to the Ranking Member, Subcommittee on Labor, Health and Human Services, Education, and Related Agencies, Committee on Appropriations, House of Representatives, Publication No. GAO-18-199, dated March 2018.

OMB	Office of Management and Budget
USDA	U.S. Department of Agriculture

WHY GAO DID THIS STUDY

Arsenic, an element in the earth's crust, can be harmful to human health and may be present in water and certain foods, such as rice. Rice may be more susceptible to arsenic contamination than other crops due to the flooded conditions in which it is typically grown. FDA and USDA work to address food safety risks. FDA's responsibilities for rice include regulatory and research programs; USDA's include research programs.

GAO was asked to review issues related to arsenic and rice. GAO examined (1) what NRC and recent key scientific reviews have reported about the effects of ingestion of arsenic on human health, (2) the extent to which FDA and USDA have managed the risk to human health from arsenic in rice, and (3) the extent to which FDA has coordinated with USDA and other federal agencies on actions to manage the risk. GAO analyzed a 2013 NRC report on inorganic arsenic, 14 reviews of scientific studies on the human health effects of ingesting arsenic published from January 2015 to June 2017, and agency documents; interviewed agency officials; and compared good practices with actions FDA and USDA took to manage risk and that FDA took to coordinate.

WHAT GAO RECOMMENDS

GAO is making five recommendations, including that FDA develop a timeline for updating its risk assessment and finalizing its draft guidance and that FDA and USDA develop a coordination mechanism for developing methods to detect foodborne contaminants, including arsenic. FDA and USDA generally agreed with the recommendations.

WHAT GAO FOUND

The National Research Council (NRC) of the National Academy of Sciences, in 2013, and more recent key scientific reviews reported evidence of associations between long-term ingestion of arsenic and adverse human health effects, such as cardiovascular disease. Many of the studies NRC reviewed as part of its survey of the scientific literature examined the ingestion of arsenic in drinking water, but others looked at arsenic from all sources, including dietary sources such as rice. NRC stated that evidence suggests that food, particularly rice, may be a significant source of inorganic arsenic, the more toxic of the two forms of arsenic; however, consumption of rice and levels of arsenic in rice vary widely, making it difficult to estimate arsenic intake from rice. NRC identified stronger evidence for some health effects at higher levels of arsenic—defined by NRC as 100 parts per billion or higher in drinking water—than at lower levels, which are more common in the United States, and noted that research on the health effects of ingesting lower levels of arsenic is ongoing.

The Food and Drug Administration (FDA) and the U.S. Department of Agriculture (USDA) have taken actions to manage the risk of arsenic in rice to human health, including assessing the type and prevalence of health effects that may result from long-term ingestion of arsenic in rice. FDA also has taken action to publicly communicate and report on the risk. In 2016, FDA issued a risk assessment about the human health effects from long-term ingestion of arsenic in rice and draft guidance recommending industry not exceed a level of 100 parts per billion of inorganic arsenic in infant rice cereal. FDA noted it issued this guidance because infants face a higher risk owing to their less-varied diets. However, FDA has not updated the risk assessment, which was informed by a review of scientific studies published before February 2015, or finalized the draft guidance. In prior work, GAO has found that sharing risk information and incorporating stakeholder feedback can help organizations identify and better manage risks, as well as increase transparency and accountability to Congress and taxpayers. FDA officials stated that they may update the risk assessment based on newly-available information and consider public comments before finalizing the draft guidance. However, FDA officials could not provide a specific timeline for either. By developing such a timeline, FDA could help clarify when it will take action and improve the transparency of its decisions.

FDA coordinated with USDA and other federal agencies on actions to manage the risk of arsenic in rice to varying extents. For example, FDA and USDA coordinated on developing arsenic detection methods for rice to a limited extent, although both agencies have crosscutting strategic goals for developing detection methods for foodborne contaminants, including arsenic. GAO has noted in prior work that developing interagency mechanisms to coordinate crosscutting issues may reduce potentially duplicative efforts. FDA and USDA officials stated that they coordinated on an informal basis but have no mechanism for coordinating more formally. By developing a coordination mechanism, FDA and USDA could enhance their ability to use their resources efficiently and avoid potentially duplicative efforts.

March 16, 2018

The Honorable Rosa L. DeLauro
Ranking Member
Subcommittee on Labor, Health and Human Services,
Education, and Related Agencies
Committee on Appropriations
House of Representatives

Dear Ms. DeLauro:

Arsenic, an element in the earth's crust, is present in the environment, including water, air, and soil, because of both natural occurrence and human activity. It exists in two general forms, organic and inorganic.[1] Both can be harmful to human health, but inorganic arsenic is generally

[1] The organic form of arsenic consists of arsenic compounds that include carbon. The inorganic form of arsenic consists of arsenic compounds that include elements other than carbon.

considered to be the more toxic form.[2] Inorganic arsenic has been classified as carcinogenic to humans by the International Agency for Research on Cancer.[3] According to the Department of Health and Human Services' (HHS) Agency for Toxic Substances and Disease Registry, exposure to arsenic generally occurs through contaminated groundwater and the ingestion of foods containing arsenic compounds.[4] The Agency for Toxic Substances and Disease Registry identified rice as a predominant dietary source of arsenic.[5] Partly because of the flooded conditions under which rice is typically grown, rice may be more susceptible to arsenic contamination than other crops. In 2009, the European Food Safety Authority noted that exposure of infants to arsenic in rice products is a concern because such products are often used in foods for infants and because infants and young children have a higher food intake relative to their body weight than adults.[6]

The safety and quality of the U.S. food supply, both domestic and imported, are governed by a complex system stemming from at least 30 federal laws that are administered by 15 federal agencies. HHS's Food and Drug Administration (FDA) and the U.S. Department of Agriculture (USDA) are the principal agencies working to protect the food supply from risks, such as contamination. FDA is responsible for ensuring the safety of virtually all domestic and imported food products, with the exception of meat and poultry, processed egg products, and catfish, which are the responsibility of USDA's Food Safety and Inspection Service. Other

---

[2] Most data reported for arsenic in food describe the levels of total arsenic (the sum of both forms). In this chapter, we use the term "arsenic" to refer to either total arsenic or inorganic arsenic only.

[3] The International Agency for Research on Cancer is the World Health Organization's agency that promotes international collaboration in cancer research. For additional information on its classifications, see International Agency for Research on Cancer, Agents Classified by the IARC Monographs, Volumes 1-120, accessed November 9, 2017, http://monographs.iarc.fr/ENG/Classification/index.php.

[4] Agency for Toxic Substances and Disease Registry, *Addendum to the Toxicological Profile for Arsenic* (Atlanta, GA: 2016).

[5] The Agency for Toxic Substances and Disease Registry noted that as a dietary source of arsenic, including inorganic arsenic, rice is second to seafood, which mostly contains the organic form of arsenic.

[6] In this chapter, we use the term "rice" to encompass rice grain and products made with rice, such as infant rice cereal.

agencies within USDA also play a role in food safety. For example, USDA's Agricultural Research Service (ARS) and National Institute of Food and Agriculture (NIFA) sponsor food safety research. In addition, the Environmental Protection Agency (EPA) has the statutory responsibility for, among other things, ensuring that the pesticides used on food crops do not endanger public health. EPA has been updating its assessment of the health hazards of inorganic arsenic in the Integrated Risk Information System—an important source of toxicity information for federal agencies. The National Research Council (NRC) of the National Academy of Sciences has reviewed EPA's approach to updating its toxicological assessment of inorganic arsenic and issued a report in 2013 that provided key guidance and a preliminary survey of the literature on inorganic arsenic.[7]

For more than 4 decades, we have reported on the fragmented federal food safety oversight system, which has caused inconsistent oversight, ineffective coordination, and inefficient use of resources. In January 2007, because of risks to the economy and to public health and safety, we added transforming federal oversight of food safety to our list of areas at high risk for fraud, waste, abuse, and mismanagement, or most in need of transformation.[8] In our February 2017 update to that list, we noted that HHS, USDA, and the Office of Management and Budget (OMB) have taken some positive steps to address fragmentation in the federal food oversight system but that additional steps are needed.[9] In January 2017, to address ongoing fragmentation, we recommended that the Executive Office of the President, in consultation with relevant federal agencies and other stakeholders, develop a national strategy to improve the food safety oversight system.[10] In addition, we have found shortcomings with FDA's

[7] National Research Council, Committee on Inorganic Arsenic, *Critical Aspects of EPA's IRIS Assessment of Inorganic Arsenic: Interim Report* (Washington, D.C.: National Academies Press, 2013).

[8] GAO, *High-Risk Series: An Update*, GAO-07-310 (Washington, D.C.: Jan. 31, 2007).

[9] GAO, *High-Risk Series: Progress on Many High-Risk Areas, While Substantial Efforts Needed on Others*, GAO-17-317 (Washington, D.C.: Feb. 15, 2017).

[10] GAO, *Food Safety: A National Strategy Is Needed to Address Fragmentation in Federal Oversight*, GAO-17-74 (Washington, D.C.: Jan. 13, 2017).

oversight of seafood, dietary supplements, and other matters.[11] In 2012, we identified key issues to consider that could benefit interagency collaborative mechanisms, which may help reduce potentially duplicative, overlapping, and fragmented efforts.[12]

You asked us to review issues related to arsenic and rice. This chapter examines (1) what NRC and recent key scientific reviews have reported about the effects of ingestion of arsenic on human health, (2) the extent to which FDA and USDA have managed the risk to human health from arsenic in rice, and (3) the extent to which FDA has coordinated with USDA and other federal agencies on actions to manage the risk.

To determine what NRC and recent key scientific reviews have reported about the effects of ingestion of arsenic on human health, we analyzed NRC's 2013 report on inorganic arsenic and 14 reviews of the scientific literature published from January 2015 through early June 2017 on the human health effects of ingestion of arsenic. We identified the reviews by conducting a literature search of research databases, such as PubMed and Toxline, and selected English-language reviews that met certain criteria, such as relying on human, rather than animal, studies. We assessed the scientific and statistical credibility, reliability, and methodological soundness of the reviews and excluded reviews for which we could not clearly determine the methodology.

To determine the extent to which FDA and USDA have managed the risk to human health from arsenic in rice, we examined relevant provisions in the Federal Food, Drug, and Cosmetic Act, as amended;[13] the Federal Agriculture Improvement and Reform Act of 1996;[14] and other laws, regulations, and policies. We also reviewed our prior work on enterprise risk management and used the essential elements for managing risk that it

---

[11] See, for example, GAO, *Imported Seafood Safety: FDA and USDA Could Strengthen Efforts to Prevent Unsafe Drug Residues,* GAO-17-443 (Washington, D.C.: Sept. 15, 2017); *Dietary Supplements: FDA May Have Opportunities to Expand Its Use of Reported Health Problems to Oversee Products,* GAO-13-244 (Washington, D.C.: Mar. 18, 2013); and *Imported Food Safety: FDA's Targeting Tool Has Enhanced Screening, but Further Improvements Are Possible,* GAO-16-399 (Washington, D.C.: May 26, 2016).

[12] GAO, *Managing for Results: Key Considerations for Implementing Interagency Collaborative Mechanisms*, GAO-12-1022 (Washington, D.C.: Sept. 27, 2012).

[13] 21 U.S.C. §§ 301, et seq.

[14] Pub. L. No. 104-127, 110 Stat. 888 (Apr. 4, 1996).

identified.[15] We identified information on agency actions for managing the risk from arsenic in rice by collecting documentation and interviewing officials from FDA and USDA, and we reviewed the information in light of the requirements, policies, and elements. We assessed FDA's and USDA's reported actions to determine the extent to which each agency's actions aligned with the essential elements for managing risk. In assessing FDA's and USDA's actions against these essential elements, we used the terms "consistent" and "partially consistent" to reflect the extent to which each agency's actions aligned with an essential element. We also interviewed 17 stakeholders, including university researchers, representatives of a consumer organization, and representatives of the rice industry, including rice mills and farms, to obtain their views on the extent to which FDA's and USDA's actions managed this risk. We identified stakeholders based on suggestions from agency officials and other stakeholders; through our site visit to Arkansas' rice agricultural research and production areas and rice mills; and the stakeholders' unique perspective or qualifications, such as membership in the NRC Committee on Inorganic Arsenic. The views we obtained in these interviews are not generalizable to all university researchers or consumer or rice industry organizations but they provide illustrative examples of the views of such stakeholders.

To determine the extent to which FDA coordinated with USDA and other federal agencies on actions to manage the risk from arsenic in rice, we identified actions for which coordination would be expected. Specifically, we identified actions for which the agencies shared similar goals in their strategic plans or relevant expertise. For these actions, we examined whether FDA developed interagency collaborative mechanisms. We had previously reported that such mechanisms could facilitate coordination between agencies.[16] We also examined whether FDA considered a key issue we had identified when implementing one of these

---

[15] GAO, *Enterprise Risk Management: Selected Agencies' Experiences Illustrate Good Practices in Managing Risk*, GAO-17-63 (Washington, D.C.: Dec. 1, 2016). This chapter also identified good practices agencies are implementing that illustrate the essential elements of enterprise risk management.

[16] GAO-12-1022.

mechanisms: to clarify the roles and responsibilities of participating agencies. We selected this key issue because it was relevant to the challenges FDA and the other agencies faced. See appendix I for a detailed description of our objectives, scope, and methodology.

We conducted this performance audit from December 2016 to March 2018 in accordance with generally accepted government auditing standards. Those standards require that we plan and perform the audit to obtain sufficient, appropriate evidence to provide a reasonable basis for our findings and conclusions based on our audit objectives. We believe that the evidence obtained provides a reasonable basis for our findings and conclusions based on our audit objectives.

BACKGROUND

Forms and Distribution of Arsenic

Arsenic is a naturally occurring element that is widely distributed in the earth's crust in two general forms—organic and inorganic. It commonly enters the body through ingestion of food or water. Most data reported for arsenic in food describe the levels of total arsenic because analyses that provide information about the forms of arsenic present are more difficult to perform, and relatively few laboratories are able to perform these analyses. Data on the levels of specific forms of arsenic, however, are becoming increasingly important because, according to the Agency for Toxic Substances and Disease Registry, the two forms have different toxicities, with inorganic arsenic being considered the more toxic form. Further, foods may have different proportions of organic and inorganic arsenic as well as different levels of total arsenic. According to the European Food Safety Authority, plants generally contain low levels of both total and inorganic arsenic, but rice may contain significant levels of total arsenic and inorganic arsenic. Levels of arsenic in groundwater, a major source of drinking water in many parts of the world, may be high in some areas; essentially all the arsenic in drinking water is inorganic arsenic.

The form and level of arsenic in rice may vary depending on the geographic region where rice is grown, conditions under which rice is grown, variety of the rice, and rice milling practices.[17] In the United States—where, according to USDA, approximately 80 percent of the rice consumed domestically is grown—rice is primarily grown in six states: Arkansas, California, Louisiana, Mississippi, Missouri, and Texas. In 2016, the latest year for which USDA data were available, about 47 percent of the rice grown in the United States was grown in Arkansas, and about 21 percent was grown in California.[18]

The amount of arsenic rice absorbs varies by geographic region because of differing levels of arsenic in the soil and other factors. Arsenic levels in the soil vary both naturally and as a result of human activity. Natural processes that contribute to arsenic levels in the soil may include bedrock weathering, because arsenic is present in many rock-forming minerals. Human activities that contribute to arsenic levels in the soil may include the use of arsenic-based pesticides and animal drugs, the mining and smelting of metal, and coal combustion.[19] Figure 1 shows the results of a 2013 U.S. Geological Survey sampling of soils to measure the levels of arsenic in the contiguous United States.[20] In addition, the figure shows the outlines of rice-growing counties based on 2016 data from USDA.

---

[17] Milling is the extent to which the hull, bran, and germ have been removed from the rice seed. White rice results from the removal of the hull and practically all of the germ and the bran layer.

[18] According to our analysis of USDA data, shares of domestic rice production in 2016 for the remaining four states were: Louisiana, 13 percent; Missouri, 7 percent; and Mississippi and Texas, 6 percent each.

[19] One arsenic-based pesticide, monosodium methanearsonate, is currently permitted for use on cotton and turf grass (i.e., golf courses, sod farms, and highway rights-of-way). In December 2015, FDA issued a final rule withdrawing the approval of the last remaining arsenic-containing animal drug approved for use in food-producing animals. 80 Fed. Reg. 78,970 (Dec. 18, 2015).

[20] U.S. Geological Survey, *Geochemical and Mineralogical Data for Soils of the Conterminous United States* (Reston, VA: 2013). The 2013 data are the most recent available.

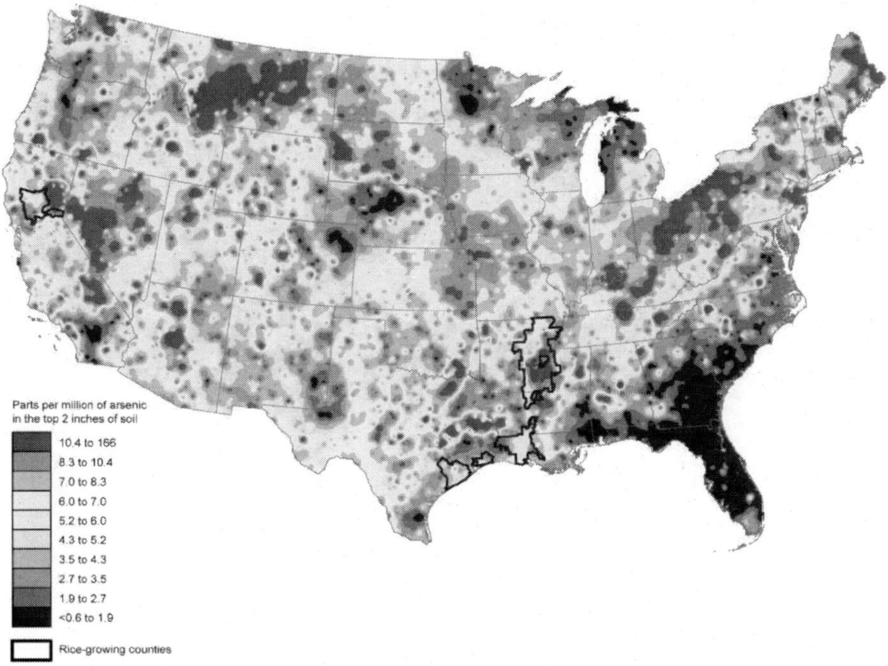

Source: Map from U.S. Geological Survey and GAO Analysis of USDA data. | GAO-18-199.

Note: Rice-growing counties are identified by USDA's National Agricultural Statistics Service as the counties within the major rice-producing regions of the United States.

Figure 1. Levels of Arsenic in the Soil Surface of the Contiguous United States, with Rice-Growing Counties Outlined.

Compared to other plants, rice absorbs more arsenic from the environment, in part because of the physiology of rice. For example, rice may readily absorb certain compounds of arsenic because, among other reasons, these compounds are similar in size to compounds containing silicon, an essential nutrient for rice. The conditions under which rice is grown may also cause it to absorb more arsenic than other plants. For instance, rice is often grown in flooded fields to control pests, grasses, and diseases, among other reasons. However, flooded conditions may promote the formation of arsenic compounds that may be easily absorbed by the rice plant. Even under the same growing conditions, some varieties of rice tend to have higher levels of arsenic in their grain, on average, than others, owing to a need for longer growing periods, among other factors. In addition, the concentrations of the two forms of arsenic may vary within the rice grain. While organic arsenic may be distributed throughout the rice grain, most of the inorganic arsenic is found in the bran layer. As seen in Figure 2, the process of milling rice removes the bran layer; thus, levels of inorganic arsenic in white, or milled, rice may be lower than those in brown, or whole grain, rice.

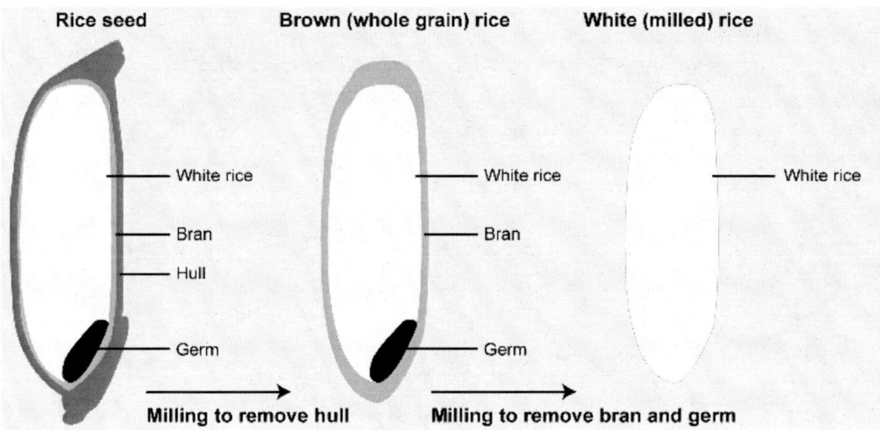

Source: GAO adaptation of image from Riceland Foods. | GAO-18-199.

Figure 2. Rice Grain Before and After Milling.

Federal Agencies' Responsibilities for Rice

A number of federal agencies are responsible for ensuring the safety and quality of rice and for assessing the human health effects of ingestion of arsenic in rice. Within HHS, FDA has overall responsibility for implementing provisions of the Federal Food, Drug, and Cosmetic Act, as amended. Specifically, FDA is responsible for determining whether food, including rice, is deemed to be adulterated (i.e., whether it bears or contains any poisonous or deleterious substance that may render it injurious to health). Under its regulations, FDA may issue guidance to establish a level of a contaminant that a food should not exceed.[21] FDA would consider case-by-case whether a food that contains the contaminant is adulterated. For example, in 2013, FDA issued draft guidance for arsenic in apple juice, on the basis of its risk assessment that estimated the long-term cancer risk posed by inorganic arsenic. According to FDA, its Center for Food Safety and Applied Nutrition is responsible for regulatory and research programs that address the health risks associated with foodborne contaminants and is aided in this role by the Office of Regulatory Affairs, which is responsible for field-based activities such as inspections, sampling, and testing of regulated products. The Center for Food Safety and Applied Nutrition also conducts industry outreach and educates consumers, among other things.

Other agencies within HHS may also conduct research, collect data, and provide information on the health effects of arsenic. For example, the National Institutes of Health (NIH) sponsor research on the health effects of ingestion of arsenic. The Centers for Disease Control and Prevention (CDC) administer the National Health and Nutrition Examination Survey, which, among other things, collects data about diet and exposure to certain substances, such as arsenic. Under the Superfund Amendments and Reauthorization Act of 1986,[22] the Agency for Toxic Substances and Disease Registry prepares toxicological profiles for certain hazardous substances, including arsenic.[23]

Agencies within USDA conduct and sponsor research to advance food safety and to help farmers market rice and manage the risk of growing it. Within USDA, ARS and NIFA conduct and sponsor research, to, among other things, maintain an adequate, nutritious, and safe supply of food to meet human nutritional needs and requirements. NIFA also distributes

---

[21] FDA may establish such a level, referred to as an "action level," when a contaminant cannot be avoided by good manufacturing practices and a technological or other change may be foreseeable. Such change may include anticipated improvements in good manufacturing practices or studies that may provide significant new toxicological or use data.

[22] 42 U.S.C. § 9604 (i)(3).

[23] In 2007, the Agency for Toxic Substances and Disease Registry issued a peer-reviewed toxicological profile for arsenic and information regarding levels of significant human exposure and significant health effects. In 2016, it issued an addendum, providing a non-peer reviewed supplement of the scientific data that were published since the release of the profile in 2007.

capacity grants that support research and extension programs at land-grant universities, which provide science-based information to farmers. The Agricultural Marketing Act of 1946 authorizes the Federal Grain Inspection Service (FGIS) to establish quality standards,[24] including standards for rice.[25] FGIS also offers inspection services for rice farmers and processors upon request. The Risk Management Agency manages the Federal Crop Insurance Corporation, which offers crop insurance to farmers for over 100 different crops, including rice. For the 2018 crop year, the rice crop insurance provisions generally require that the rice be flood-irrigated (i.e., intentionally covered with water at a uniform and shallow depth throughout the growing season).[26]

Other agencies play a role in managing the risk of arsenic. EPA regulates the presence of certain substances, such as arsenic, in drinking water under the Safe Drinking Water Act and conducts toxicological assessments.[27] In 2001, EPA issued a rule limiting the level of arsenic in drinking water to 10 parts per billion (ppb) to protect consumers from the health effects of long-term exposure.[28] Under its Integrated Risk Information System program, EPA conducts assessments that provide toxicity values—such as for increased cancer risk due to lifetime ingestion of a specified quantity of a substance. In accordance with congressional direction,[29] EPA submitted a plan for developing a draft assessment and preliminary assessment materials for inorganic arsenic to NRC for review. In 2013, NRC released an interim report, which provided guidance to EPA and included a preliminary survey of the scientific literature.[30] In addition, in accordance with Executive Order 13272, the Small Business

---

[24] 7 U.S.C. § 1622(h)(1).

[25] Federal Grain Inspection Service, *United States Standards for Rice* (Washington, D.C.: 2009).

[26] The Federal Crop Insurance Corporation also offers a sprinkler irrigated endorsement that, if purchased, allows farmers to irrigate rice with sprinklers rather than with continuous flooding.

[27] 42 U.S.C. § 300g–1(b)(12)(A).

[28] 66 Fed. Reg. 6976 (Jan. 22, 2001). EPA defines chronic or long-term exposure as multiple exposures occurring over a significant portion of a lifetime.

[29] H. Conf. Rpt. 112-331 (2012).

[30] NRC has also previously issued reports in 1999 and 2001 to help evaluate the scientific validity of the risk assessment that EPA used in its rulemaking limiting arsenic in drinking water.

Administration's Office of Advocacy helps agencies assess the potential impacts of draft rules on small businesses—which could include members of the rice industry—small governmental jurisdictions, and small organizations.[31]

Entities outside of the federal government have recently proposed or established limits or guidance for arsenic in rice. For example, in 2017, the Codex Alimentarius, an international standard-setting body, published a code of practice that provides guidance for preventing and reducing arsenic contamination in rice, as well as communicating the risk to stakeholders.[32] In 2014 and 2016, the Codex Alimentarius established a standard for inorganic arsenic of 200 ppb for white rice and 350 ppb for brown rice.[33] In 2015, the European Commission issued a regulation limiting inorganic arsenic in various rice-based foods, including limits of 200 ppb in white rice, 250 ppb in brown rice, and 100 ppb in rice destined for food for infants and young children.[34]

Enterprise Risk Management

Enterprise risk management allows agencies to assess threats and opportunities that could affect the achievement of their goals. In a 2016 report, we updated our 2005 risk management framework to (1) reflect changes to OMB's Circular A-123, which requires agencies to implement enterprise risk management;[35] (2) incorporate recent federal experience;

---

[31] Executive Order 13272, *Proper Consideration of Small Entities in Agency Rulemaking* (Washington, D.C.: Aug. 13, 2002).

[32] Codex Alimentarius, *Code of Practice for the Prevention and Reduction of Arsenic Contamination in Rice*, CAC/RCP 77-2017.

[33] Codex Alimentarius, *General Standard for Contaminants and Toxins in Food and Feed*, Codex Standard 193-1995. The Codex Alimentarius standards are voluntary for nations but, according to the World Trade Organization, they are used as a benchmark for trade purposes. The World Trade Organization encourages reference to the standards in bilateral and multilateral trade agreements. Various federal agencies, including EPA, FDA, and USDA, participate in Codex Alimentarius activities.

[34] Commission Regulation (EU) 2015/1006 of 25 June 2015, amending Regulation (EC) No 1881/2006 as regards maximum levels of inorganic arsenic in foodstuffs.

[35] Office of Management and Budget, *Management's Responsibility for Enterprise Risk Management and Internal Control,* OMB Circular No. A-123 (July 15, 2016).

and (3) identify essential elements of federal enterprise risk management.[36] Beyond traditional internal controls, enterprise risk management promotes risk management by considering its effect across the entire organization and how it may interact with other identified risks. Additionally, it also addresses other topics such as setting strategy, governance, communicating with stakeholders, and measuring performance, and its principles apply at all levels of the organization and across all functions—such as those related to managing the risk of arsenic in rice. The six essential elements of enterprise risk management that we identified in December 2016 are as follows:[37]

- Align risk management process with goals and objectives. Ensure the process maximizes the achievement of agency mission and results.
- Identify risks. Assemble a comprehensive list of risks, both threats and opportunities, that could affect the agency's ability to achieve its goals and objectives.
- Assess risks. Examine risks, considering both the likelihood of the risk and the impact of the risk to help prioritize risk response.
- Respond to the risks. Select risk treatment response (based on risk appetite), including acceptance, avoidance, reduction, sharing, or transfer.
- Monitor risks. Monitor how risks are changing and whether responses are successful.
- Communicate and report on risks. Communicate risks with stakeholders and report on the status of addressing the risks.

---

[36] GAO-17-63. See also GAO, *Risk Management: Further Refinements Needed to Assess Risks and Prioritize Protective Measures at Ports and Critical Infrastructure,* GAO-06-91 (Washington, D.C.: Dec. 15, 2005).

[37] In this chapter, we used "risk management" rather than "enterprise risk management" because we examined FDA's and USDA's actions related to arsenic and rice, rather than their entire portfolio of activities related to food safety.

NRC and Recent Key Scientific Reviews Reported Evidence of Associations between Ingestion of Arsenic and Adverse Human Health Effects

NRC, in its 2013 report, and recent key scientific reviews reported evidence of associations between long-term ingestion of arsenic and adverse human health effects.[38] NRC identified stronger evidence of these associations at higher arsenic levels—defined by NRC as 100 ppb or higher in drinking water—than at lower levels, which are more common in the United States. NRC reported greater uncertainty regarding the associations with some health effects at lower levels of arsenic and noted that research on the health effects of ingestion of lower levels of arsenic is ongoing. Many of the studies on which NRC based its conclusions were focused on the ingestion of arsenic from drinking water, but other studies were based on arsenic from all sources, including dietary sources such as rice.[39] Further, NRC reported that evidence from CDC dietary surveys and related academic studies suggests that food, particularly rice, may be a significant source of inorganic arsenic, especially when arsenic levels in drinking water are lower; however, consumption of rice and levels of arsenic in rice vary widely, making it difficult to estimate arsenic intake from rice. NRC reported strong evidence of causal associations—that is, a

---

[38] Studies report an association between exposure—for example, through ingestion—to a substance and a health effect if the occurrence of the health effect is significantly different between groups exposed to different levels of the substance. An association, however, does not necessarily mean that the exposure caused the health effect, but according to NRC, associations can help identify the hazards. The NRC report categorized health effects as follows: (1) strong evidence of causal association, (2) moderate evidence of association, and (3) limited but suggestive evidence of association. To categorize a health effect, according to the chair of the NRC committee that developed the report, the committee relied on nine factors such as strength and consistency. Health effects in the first category met all factors, those in the second met many of them, and those in the third met a few.

[39] NRC reviewed 257 studies as part of its preliminary review of the scientific literature on the epidemiologic evidence related to the most affected organ systems. In certain studies, researchers estimated the ingested level of arsenic based on levels of arsenic measured in urine, hair, or toenails. In such cases, the arsenic may have originated in water or food. According to NRC, after ingestion, inorganic arsenic in drinking water is rapidly and almost completely absorbed, and there is evidence that arsenic in rice is likewise almost completely absorbed. We did not examine the individual studies cited by the NRC report.

potential cause and effect—between the long-term ingestion of arsenic from water or dietary sources, such as rice, and the following five health effects:

Skin Diseases

Skin Lesions

Skin lesions due to arsenic ingestion predispose a person to some skin cancers and may indicate increased susceptibility to other cancer and noncancer diseases. Skin lesions have a well-established dose-response relationship with arsenic in drinking water.[40]

Skin Cancer

Arsenic is an established skin carcinogen, according to NRC. NRC stated that almost all published studies found evidence of an association between arsenic ingestion and nonmelanoma skin cancers.[41]

Lung Cancer

Arsenic from drinking water is an established lung carcinogen in humans, according to NRC. NRC cited studies conducted in Argentina, Chile, Japan, Taiwan, and the United States that reported associations between high levels of arsenic ingestion and lung cancer. NRC reviewed several studies that examined ingestion of lower levels of arsenic, some of which found evidence of an association, while others did not.[42]

---

[40] A dose-response relationship, in this case, describes how the likelihood and severity of an adverse health effect (the response) is related to the amount and type of exposure to arsenic (the dose). NRC reported that, while some studies found that skin lesions occur only at high levels of arsenic, others found that lesions also occur at lower levels—that is, below 100 ppb of arsenic in drinking water.

[41] Some studies have found evidence that the risk of skin cancer increases with increasing levels of arsenic. Studies that examined ingestion of lower arsenic levels have generally been less conclusive, in part because of methodological limitations.

[42] One study in Japan found evidence of an increase in lung cancer incidence with increased consumption of foods containing arsenic. In addition, one study in Chile suggests that ingestion of arsenic in utero or in early life increases the risk of lung cancer mortality, and a

Cardiovascular Disease

NRC stated that many studies found a causal association between the ingestion of arsenic and cardiovascular disease and mortality. Studies suggest that the ingestion of lower levels of arsenic in drinking water and possibly in food is associated with cardiovascular disease, but additional evidence is needed to fully understand the relationship.

Bladder Cancer

Arsenic is an established bladder carcinogen in humans, according to NRC. NRC cited a 2012 assessment by the International Agency for Research on Cancer that indicated higher mortality from bladder cancer in populations that are exposed to high levels of arsenic compared to those that are not based on studies in Argentina, Chile, and Taiwan.[43]

NRC reported that there was moderate evidence of association between the long-term ingestion of various levels of arsenic from water or dietary sources such as rice, and adverse health effects, although some studies found evidence of an association and others did not. Adverse health effects include, for example, neurodevelopmental toxicity and pregnancy outcomes related to infant illness, disease, or injury.[44] NRC also reported that there was limited evidence of an association between the long-term ingestion of arsenic from water and dietary sources and adverse health effects, such as liver and pancreatic cancer and renal disease.

We analyzed 14 scientific reviews, published since NRC's 2013 report, from January 2015 through early June 2017, that generally have supported NRC's conclusions that long-term ingestion of arsenic is associated with the above-mentioned health effects. Two reviews reporting additional evidence related to cardiovascular disease suggested that there may be a

---

study in the United States found a relationship between increased urinary arsenic levels and lung cancer mortality.

[43] Officials from the National Cancer Institute provided us with a 2016 study that found a dose-response relationship between bladder cancer and ingestion of drinking water with lower levels of arsenic in the United States.

[44] NRC reported that there is increasing evidence that exposure to arsenic early in life, including during the prenatal and early childhood stages, when rice products are often ingested, can affect health early or later in life. During these more sensitive life stages, exposure may not necessarily be long-term.

threshold—an arsenic level below which there is no significant occurrence of cardiovascular disease.[45] However, one of these reviews noted that the number of studies they examined was small, among other limitations. Regarding lung cancer, another recent review proposed a dose-response relationship, which NRC identified as a gap in the understanding of this adverse health effect.[46] However, this review noted that the studies it included did not distinguish between the risk of lung cancer in smokers and non-smokers, which NRC reported may be a key confounding factor. The review also cited other limitations, including the small number of studies it used to model this relationship. See appendix II for additional information about the reviews we identified.

FDA and USDA Have Taken Actions to Manage the Risk to Human Health from Arsenic in Rice

FDA and USDA have taken actions to manage the risk to human health from arsenic in rice, including assessing the type and prevalence of health effects that may result from long-term ingestion. These efforts were generally consistent with the six essential elements for managing risk, which we have found could help agencies assess threats that could affect the achievement of their goals. Specifically, FDA has taken actions that were consistent with five of the six essential elements, including: (1) aligning risk management process with goals and objectives, (2) identifying risks, (3) assessing risks, (4) responding to the risks, and (5) monitoring risks. However, FDA has not fully taken action on the sixth element of communicating and reporting on risks. FDA issued a risk

[45] D. Phung, D. Connell, S. Rutherford, and C. Chu. "Cardiovascular Risk from Water Arsenic Exposure in Vietnam: Application of Systematic Review and Meta-Regression Analysis in Chemical Health Risk Assessment," *Chemosphere*, vol. 177 (2017); and M.S. Sidhu, K.P. Desai, H.N. Lynch, L.R. Rhomberg, B.D. Beck, and F.J. Venditti. "Mechanisms of Action for Arsenic in Cardiovascular Toxicity and Implications for Risk Assessment," *Toxicology*, vol. 331 (2015).

[46] S.H. Lamm, H. Ferdosi, E.K. Dissen, J. Li, and J. Ahn, "A Systematic Review and Meta-Regression Analysis of Lung Cancer Risk and Inorganic Arsenic in Drinking Water," *International Journal of Environmental Research and Public Health*, vol. 12, no. 12 (2015).

assessment in 2016 for public comment and a draft guidance limiting the levels of arsenic in infant rice cereal, but it has not updated or finalized these key documents. USDA has taken actions consistent with five of the six essential elements but has not taken actions to monitor the risk because of its more limited, nonregulatory role.

Aligning Risk Management Process with Goals and Objectives

FDA and USDA have aligned their actions to manage the risk to human health from arsenic in rice to goals in their strategic plans. According to FDA officials, FDA's actions align with three of the six goals identified in the 2015–2018 research strategic plan for FDA's Center for Food Safety and Applied Nutrition, including advancing diet and health research that contributes to the development of science-based policies and communication strategies.[47] Regarding USDA's actions, ARS officials stated that their research on arsenic in rice aligned with four goals in ARS's fiscal year 2012–2017 strategic plan, such as protecting food from pathogens, toxins, and chemical contamination during production, processing, and preparation.[48] NIFA officials stated that the research they sponsored on arsenic in rice aligned with one of the sub-goals in NIFA's fiscal year 2014–2018 strategic plan: to reduce the incidence of foodborne

---

[47] Center for Food Safety and Applied Nutrition, *Science and Research Strategic Plan, 2015–2018.* These goals are as follows: Goal 2, develop and implement screening methods for use in field laboratories to improve the capacity for detection of chemical contaminants; Goal 4, integrate and apply modern toxicological approaches to support regulatory and public health decision making on chemical hazards in foods, dietary supplements, and cosmetics; and Goal 5, advance diet and health research that contributes to the development of science-based policies and communication strategies.

[48] Agricultural Research Service, *Strategic Plan FY 2012–FY 2017* (Washington, D.C.: 2016). These goals are as follows: Goal 1.1, enable Americans to make health-promoting, science-based dietary choices; Goal 1.2, protect food from pathogens, toxins, and chemical contamination during production, processing, and preparation; Goal 2.1, integrated, effective, and safe water resource management; and Goal 3.1, protect, expand, and enhance the United States' crop genetic resource base, increase scientific knowledge of crop genes, genomes, biological processes and systems, and deliver economically and environmentally sound technologies that improve the production efficiency, quality, health and value of the nation's crops.

illness and provide a safer food supply.[49] FGIS officials provided doumentation showing that their actions aligned with one of the goals in their fiscal year 2016–2020 strategic plan: provide the environment for fair and competitive market practices between agricultural producers and buyers.[50] FDA's and USDA's actions were consistent with the essential element of aligning risk management actions to their strategic plans.

Identifying Risks

FDA and USDA have taken actions to identify the risk of arsenic in rice. FDA has identified the risk of arsenic in rice through the Total Diet Study—an annual testing of contaminants and nutrients in food. As part of conducting the Total Diet Study, FDA collects samples of certain foods, including rice, and tests them for a variety of toxic chemicals, including total arsenic. From 2014 through 2015, the most recent years for which data are available, FDA tested six different categories of rice-based foods for arsenic.[51] FDA officials told us that they identified arsenic in rice as a priority based, in part, on the results of the Total Diet Study, which indicated that rice had higher levels of arsenic compared to other foods. Some university researchers we interviewed stated that the Total Diet Study would be more helpful if it measured inorganic arsenic or had a more robust methodology. For example, one university researcher noted that the number of samples in the Total Diet Study is not big enough to be nationally representative. FDA officials told us that starting with the fiscal year 2018 Total Diet Study, they plan to begin testing rice-based foods for inorganic arsenic, increase the number of samples they collect, and make other improvements to the sampling methodology.

---

[49] National Institute of Food and Agriculture, *Strategic Plan, FY 2014–FY 2018* (2014).

[50] Grain Inspection, Packers and Stockyards Administration, *Strategic Plan, FY 2016- 2020.* In September 2017, the programs carried out by FGIS, formerly part of the Grain Inspection, Packers and Stockyards Administration, were realigned into USDA's Agricultural Marketing Service.

[51] FDA included three baby foods—turkey and rice; chicken and rice; and dry rice cereal, prepared with water—as well as cooked white enriched rice, crisped rice cereal, and meatless fried rice from restaurants that serve Chinese carry-out food.

USDA officials have taken actions to identify the risk of arsenic in rice through a variety of research programs. ARS officials told us that they have conducted research on arsenic in rice under four national programs: (1) plant genetic resources, genomics, and genetic improvement; (2) water availability and watershed management; (3) human nutrition; and (4) food safety. For example, ARS researchers are examining whether changes in soil chemistry as a result of organic or conventional management practices affect arsenic levels in rice. NIFA officials stated that NIFA sponsors research on arsenic in rice through formula-based grants to universities and through competitive grants, such as those offered through the Agriculture and Food Research Initiative.[52]

Total Diet Study

The Food and Drug Administration's (FDA) Total Diet Study, which began testing for arsenic in 1991, is an ongoing program that monitors the levels of about 800 contaminants and nutrients in the average U.S. diet. To conduct the study, FDA buys, prepares, and analyzes about 280 kinds of foods and beverages from representative areas of the country and estimates the average amounts of contaminants and nutrients the entire U.S. population, some subpopulations, and each person consumes annually. The sampling plan calls for purchasing each type of food four times a year, each time in a different region. Within each region, FDA purchases each food product from three different stores and combines them into a composite sample, for a total of four estimates each year. FDA makes results of the study, from 1991 through 2015, available to the public in electronic form on its website.

Source: GAO analysis of FDA data. | GAO-18-199.

To identify what research to undertake, ARS officials told us that they typically meet with industry to identify its highest priorities. For example, ARS officials from the Delta Water Management Research Unit in Arkansas stated that they started researching arsenic in rice after participating in a joint ARS-USA Rice Federation conference in 2012.[53]

[52] The Agriculture and Food Research Initiative is a competitive grant program that supports research, education, and extension activities in six priority areas. They include (1) plant health and production and plant products; (2) animal health and production and animal products; (3) food safety, nutrition, and health; (4) bioenergy, natural resources, and environment; (5) agriculture systems and technology; and (6) agriculture economics and rural communities.

[53] USA Rice Federation is an organization that advocates for the U.S. rice industry, including rice farmers and millers.

FGIS officials told us that contaminants such as arsenic may affect the quality of a grain, such as rice, and hence its value. They stated that they work closely with the grain industry to develop new standards and tests to meet industry's needs.

Assessing Risks

FDA and USDA have taken actions to assess the risk of arsenic in rice. In 2012, FDA published its current method to detect inorganic arsenic in rice.[54] FDA officials told us that this method, though useful, is time-consuming and expensive, and the agency continues to develop other methods to reduce cost and time. For example, in 2017, FDA developed another method to detect inorganic arsenic in wine and rice that takes less time than its current method.[55] FDA officials told us they have an ongoing research project on a field-deployable method based on a commercially-available digital arsenic test kit for detecting arsenic in drinking water called the Arsenator. In addition, FDA has been using laser ablation, the process of removing a material from a solid using a laser beam so that it can be measured, as a way to study arsenic distribution in rice.[56]

From 2011 through 2014, FDA conducted targeted sampling of more than 1,400 rice-based foods—including rice, rice beverages, cereals, and snacks—for inorganic arsenic. This targeted sampling and a literature review of articles published before February 2015 informed a risk assessment of arsenic in rice that FDA issued for public comment in April 2016.[57] Specifically, the risk assessment used the results of the targeted

---

[54] Food and Drug Administration, *Arsenic Speciation in Rice and Rice Products Using High Performance Liquid Chromatography-Inductively Coupled Plasma-Mass Spectrometric Determination,* Elemental Analysis Manual for Food and Related Products, 4.11 (2012).

[55] Gray, Patrick, J., Courtney K. Tanabe, Susan E. Ebeler, and Jenny Nelson, "A Fast and Fit-for-Purpose Arsenic Speciation Method for Wine and Rice," *Journal of Analytical Atomic Spectrometry,* vol. 32 (2017).

[56] Laser ablation is used with an analytical instrument that can measure the material, such as an inductively-coupled plasma mass spectrometer.

[57] Food and Drug Administration, Center for Food Safety and Applied Nutrition, *Arsenic in Rice and Rice Products Risk Assessment Report, Version Released for Public Comment* (March 2016). 81 Fed. Reg. 19,976 (Apr. 6, 2016). FDA conducted a literature review on arsenic

sampling to identify levels of inorganic arsenic in rice and examined available scientific information to provide quantitative estimates of lung and bladder cancer risk—that is, the number of expected lung and bladder cancer cases per million people that may be attributable to longterm ingestion of inorganic arsenic in rice and a qualitative assessment of other adverse health effects. The risk assessment also analyzed alternative approaches to reducing the risk of arsenic in rice, such as instituting limits on the allowable level of arsenic in various rice-based foods, limiting the amount and frequency of consumption of rice, and cooking practices.

FDA's actions have helped assess the risk of arsenic in rice, although some stakeholders we interviewed have identified limitations to FDA's actions. For example, one rice producer noted that because FDA's current detection method is time-consuming and expensive, it is not widely used—companies only use it when tests for total arsenic reveal that the levels exceed the limit for inorganic arsenic that their customers request. Some stakeholders noted that the evidence FDA used to assess the risk of the ingestion of low levels of arsenic, which may be more relevant for rice consumption, is more uncertain.

USDA agencies have also taken actions to assess the risk by conducting research to develop faster and less expensive methods to detect inorganic arsenic in rice. In 2016, ARS developed a method using hydride generation, which uses an acid to convert the inorganic arsenic into a gas that can be detected by an instrument. ARS officials stated that they have conducted research on the hydride generation method for more than 5 years and were able to further refine the method with funding from the Rice Foundation.[58] Stakeholders from the rice industry and a university researcher we interviewed noted that, while the hydride generation method is faster and cheaper than FDA's current detection method, it is too time-consuming and expensive for commercial purposes. For example, rice mills could not keep pace with trucks lining up to unload rice if they use

---

levels in U.S. rice and infant rice cereal published up to October 2013 as well an update of articles published from October 2013 to February 2014 on cancer-related health effects and another update of articles published from October 2013 to February 2015 on noncancer health effects for pregnant women, infants, and children.

[58] The Rice Foundation sponsors research and education programs for the rice industry.

the hydride generation method. However, ARS officials stated that researchers may use it if they need to analyze thousands of samples and are willing to trade off some accuracy for speed and cost. In addition, FGIS conducted some of its own development work on the Arsenator. Agency officials said that they began research on the Arsenator to help provide a rapid and inexpensive method of detecting inorganic arsenic at FGIS official testing locations that could include rice mills but have suspended their efforts because representatives of the rice industry have told them that these tests are not necessary.

Responding to the Risks

FDA and USDA have taken actions to respond to the risk of arsenic in rice. In 2016, FDA issued draft guidance, which proposed an action level, recommending that the rice industry not exceed a level of 100 ppb inorganic arsenic in infant rice cereal, and FDA has conducted research on cooking methods that may reduce arsenic. In its draft guidance, FDA stated that it used its risk assessment, among other considerations, to identify the level of inorganic arsenic in infant rice cereal.[59] FDA further noted that it selected 100 ppb because of the potential for human health risks associated with inorganic arsenic and because such a level is achievable with the use of current good manufacturing practices— specifically, selecting sources of rice or rice-derived ingredients with lower inorganic arsenic levels.[60] FDA officials told us that they focused on infant rice cereal because infants are at a higher risk of experiencing some of the health effects of ingesting inorganic arsenic, such as neurodevelopmental effects, and because the diet of infants is less varied than that of adults. FDA officials noted that the proposed guidance sets a limit for infant rice cereal that is generally

---

[59] Food and Drug Administration, *Inorganic Arsenic in Rice Cereals for Infants: Action Level Guidance for Industry, Draft Guidance* (April 2016). 81 Fed. Reg. 19,976 (Apr. 6, 2016).

[60] A level of 100 ppb is equivalent to 100 micrograms of arsenic per kilogram of rice. In setting this level, FDA used information about the health effects of ingesting various levels of arsenic, as well as information about rice consumption patterns and levels of arsenic in rice based on its sampling efforts.

consistent with the limit set by the European Commission and that other types of rice sold in the United States also generally meet the Codex Alimentarius standards.

University researchers and a group representing consumers we interviewed stated that FDA's draft guidance is a good first step, but that FDA should establish limits for arsenic in other rice products, such as rice crackers and other foods that children eat. FDA officials noted that the next most susceptible group would likely be toddlers and young children, but because their diet is more diverse than that of infants, rice-based foods make up a smaller portion of their diet. FDA requested public comments on certain aspects of the draft guidance, such as its feasibility, and noted that when it is finalized, it will represent FDA's current thinking on this topic. The public comments were due to FDA in July 2016, although FDA noted that the public may comment on its guidance at any time. University researchers and stakeholders from the rice industry we interviewed stated that FDA's draft guidance has become a de facto industry standard for infant rice cereal. In 2016, FDA also published research on the effect that cooking methods, such as cooking rice in excess water, may have on reducing the level of arsenic in rice.[61] FDA officials told us that they provided advice to consumers on cooking methods that could reduce arsenic in rice on the FDA website but said FDA will not direct manufacturers to change the cooking instructions for rice because the alternative methods may reduce the nutritional value of the rice.[62]

Within USDA, ARS and NIFA have sponsored published and ongoing research that can help respond to the risk, such as research on ways to reduce the uptake of arsenic by rice through new rice varieties, water management practices, and soil additives, as well as research on the genetic mechanisms underlying the uptake and transport of arsenic in the rice plant.

---

[61] Gray, Patrick J., Sean D. Conklin, Todor I. Todorov, and Sasha M. Kasko, "Cooking Rice in Excess Water Reduces Both Arsenic and Enriched Vitamins in the Cooked Grain," *Food Additives and Contaminants: Part A*, vol. 33, no. 1 (2016).

[62] Food and Drug Administration, *Questions & Answers: Arsenic in Rice and Rice Products*, accessed May 19, 2017, https://www.fda.gov/Food/FoodborneIllnessContaminants/Metals/ucm319948.htm.

Flooded rice field at ARS's Dale Bumpers National Rice Research Center, Arkansas

Dry field of rice at the University of Arkansas System Division of Agriculture

Source: GAO analysis of USDA data (art); GAO (photos). | GAO-18-199.

Note: The U.S. Department of Agriculture's (USDA) Agricultural Research Service (ARS) and the University of Arkansas System Division of Agriculture, funded in part by USDA, conduct research on alternate wetting and drying, a water management practice illustrated above, to determine how to produce comparable rice yields to those from continuous flooding and reduce water use and levels of arsenic.

Figure 3. Alternate Wetting and Drying Water Management Practice in Arkansas Rice Fields.

For example, ARS has been conducting research on rice varieties that can improve yield and grain quality, including lower levels of arsenic, at the Dale Bumpers National Rice Research Center in Arkansas for more than 30 years. In 2016, university and ARS researchers published a study showing that growing rice using a water management practice called alternate wetting and drying could decrease the levels of arsenic.[63] Under

[63] LaHue, Gabriel T., Rufus L. Chaney, Maria Arlene Adviento-Borbe, Bruce A. Linquist, "Alternate Wetting and Drying in High Yielding Direct-Seeded Rice Systems Accomplishes Multiple Environmental and Agronomic Objectives," *Agriculture, Ecosystems and Environment,* vol. 229 (2016).

this practice of growing rice, shown in Figure 3 below, fields are periodically drained and re-flooded during the growing season.

ARS officials stated that the alternate wetting and drying water management practice has been adopted to a limited extent in Arkansas, but pointed out that other benefits, such as reducing water use, may have been more influential to its adoption than the lowering of arsenic levels. They noted that there are a number of challenges that may preclude widespread use, including inadequate water-pumping capacity and the lack of crop insurance coverage for the practice. In addition, in 2015, university researchers and an ARS researcher, with a grant from NIFA, published a study on the effects of adding iron oxide to the soil on the levels of arsenic in rice; they found that iron oxide resulted in significant reduction of arsenic for the two varieties of rice that the study examined.[64]

Monitoring Risks

FDA, which is responsible for ensuring the safety of rice and rice-based foods, has taken actions to monitor the risk of arsenic in rice. USDA has not done so, because of its more limited, nonregulatory role. FDA has a compliance program designed to monitor over 1,400 products annually, including foods that are most likely to contribute to the dietary intake of toxic elements, among other contaminants.[65] In fiscal years 2015 and 2016, FDA monitored the risk of arsenic by assessing the levels in rice and rice-based foods under this compliance program, and FDA officials told us that they plan to continue to do so in fiscal years 2017 and 2018.[66] FDA

---

[64] Farrow, Eric, M., Jianmin Wang, Joel G. Burken, Honglan Shi, Wengui Yan, John Yang, Bin Hua, and Baolin Deng, "Reducing Arsenic Accumulation in Rice Grain through Iron Oxide Amendment," *Ecotoxicology and Environmental Safety*, vol. 118 (2015).

[65] The Toxic Elements in Food and Foodware, and Radionuclides in Food Compliance Program focuses on sampling and analyzing foods that can be significant dietary sources of toxic elements for young children.

[66] In fiscal year 2015, FDA tested eight samples of rice-based foods for arsenic; in fiscal year 2016, FDA tested 38 samples; and FDA officials provided plans to test 93 samples in fiscal years 2017 and 86 in fiscal year 2018. The samples include infant rice cereals; rice-based infant and toddler puffs and teething biscuits; and rice beverages, including those targeted to children.

officials told us that they generally test the rice for total arsenic but have recently analyzed some samples for inorganic arsenic based on factors such as the level of total arsenic found. FDA considers whether to conduct follow-up actions, including enforcement actions, on a case-bycase basis. As a result of its monitoring in 2016 and 2017 FDA considered, but did not take, two enforcement actions for arsenic in infant rice cereal. FDA officials stated that the inorganic arsenic level in one case was close to the 100 ppb limit and within the margin of error of the detection method, and in the second case, FDA determined during its follow-up to the initial sample that the manufacturer destroyed the remaining product.

USDA agencies have not monitored arsenic in rice. The Food Safety and Inspection Service is USDA's regulatory agency for food safety, but officials have told us they have not taken actions in this area because rice is not under the agency's jurisdiction. ARS maintains a food composition database,[67] but it does not monitor rice for contaminants such as arsenic because, according to ARS officials, that is not the database's purpose. FGIS officials stated that they do not have an arsenic testing program for rice at this time. They told us that they considered establishing a testing program for rice intended for export at the request of the rice industry. However, FGIS officials stated that they suspended their efforts when industry determined that it did not need a testing program.

Communicating and Reporting on Risks

FDA and USDA have taken actions to communicate and report on the risk of arsenic in rice to the public. FDA has issued a risk assessment and draft guidance on arsenic in infant rice cereal, but it has not updated or finalized these documents. FDA's 2016 risk assessment report provides information about the risk from long-term ingestion of arsenic in rice, and its draft guidance on arsenic in infant rice cereal includes a link to an FDA

---

[67] USDA National Nutrient Database for Standard Reference contains food composition data for over 8,800 food items and components, including nutrients such as fiber, minerals, protein, and vitamins.

website with information for consumers, including pregnant women and parents.[68] FDA has requested comments and received 22 public comments from 17 individuals and organizations on both documents. The comments have addressed a range of issues, including the methodology FDA used in its risk assessment; the 100 ppb limit and scope of the agency's draft guidance; and the effectiveness of the agency's communication to the public. However, FDA has not publicly issued versions of the guidance or the risk assessment that address these comments. In our prior work, we have found that sharing risk information and incorporating feedback from internal and external stakeholders can help organizations identify and better manage risks, as well as increase transparency and accountability to Congress and taxpayers.[69]

In the risk assessment, FDA stated that it will provide an update after considering public comments and any newly-available information. For example, FDA officials told us that they plan to consider newly-available information, such as any updates to EPA's Integrated Risk Information System assessment for inorganic arsenic, and may update the risk assessment as a result. With regard to public comments, FDA officials told us that they do not intend to make any changes to the approach or findings of the risk assessment and that they are still considering whether to make changes to the draft guidance as a result of public comments. FDA officials stated that they are still reviewing comments and that, before publication, the guidance would have to undergo interagency review. FDA officials also stated that the agency is not required to provide a response to comments in the final guidance. Further, FDA officials stated that the agency does not need to finalize the guidance in order to sample foods for a contaminant or to take enforcement action when contamination may pose a health hazard.

---

[68] FDA's website provides information to consumers and FDA's statement on testing and analysis of arsenic in rice and rice products. Food and Drug Administration, "Arsenic in Rice and Rice Products," accessed May 19, 2017, https://www.fda.gov/Food/FoodborneIllnessContaminants/Metals/ucm319870.htm.

[69] GAO-17-63.

Stakeholders we interviewed stated that updating the risk assessment and finalizing the draft guidance would improve FDA's communication of the risk. For example, some stakeholders we interviewed told us that the information used in the risk assessment—both regarding the health effects of arsenic and the levels of arsenic in rice—may need to be updated to incorporate the results of more recent research. Further, two stakeholders we interviewed—one representing the rice industry and the other representing consumers—noted that it is not clear to them what actions FDA can take based on the draft guidance. However, FDA officials could not give us a timeline for when they plan to update the risk assessment or finalize the guidance. By developing a timeline for updating the risk assessment on arsenic in rice to incorporate any newly-available information, FDA could help clarify when it will take action. Developing a timeline for finalizing the draft guidance on arsenic in infant rice cereal could also help FDA improve the transparency of its decisions—such as by clarifying the effectiveness of the draft guidance.

USDA has taken actions that can help communicate and report on the risk of arsenic in rice. ARS officials told us that they have communicated the results of their research on arsenic in rice in a number of ways, such as through presentations at conferences and through outreach to farmers, including in cooperation with extension programs at universities. For example, USDA researchers demonstrated automated irrigation systems that can be used for the alternate wetting and drying water management practice. In 2017, ARS researchers contributed to the development of a bulletin in conjunction with University of Arkansas researchers that contains recommended practices about irrigation methods that can reduce the levels of arsenic in rice.[70] ARS officials told us that their communication efforts could help increase farmers' interest and adoption of methods they have researched. They also stated that they work with extension programs because these programs have good access to farmers.

---

[70] University of Arkansas Division of Agriculture Research and Extension, *Rice Information: Using Alternate Wetting and Drying (AWD) Rice Flood Management* (July 2017).

FDA Coordinated Several Risk Management Actions with USDA and Other Federal Agencies to Varying Extents

FDA coordinated with USDA and other federal agencies on the actions to manage the risk of arsenic in rice for which coordination would be expected, to varying extents. FDA coordinated with USDA and several other federal agencies, including CDC, EPA, and NIH, on the development of the risk assessment and draft guidance on arsenic in infant rice cereal, but USDA raised concerns about the extent of the coordination. FDA and USDA coordinated to a limited extent to develop faster and less expensive methods to detect arsenic in rice.

FDA Coordinated Its Risk Assessment and Draft Guidance with Several Federal Agencies, but USDA Raised Concerns about the Extent of Coordination

FDA coordinated with several federal agencies on the development of the risk assessment and draft guidance on arsenic in infant rice cereal. According to FDA officials, in developing the risk assessment, FDA initially coordinated with EPA on two noncancer health effects—adverse pregnancy outcomes and developmental neurotoxicology effects in young children—to ensure consistency with the work EPA was doing to update its Integrated Risk Information System assessment for arsenic. When FDA completed the draft of the noncancer section of its risk assessment, the agency provided it to EPA and NIH's National Institute of Environmental Health Sciences for review. FDA incorporated comments from EPA and NIH in the risk assessment document, which EPA, CDC, and NIH subsequently reviewed. From December 2014 through June 2015, the risk assessment and draft guidance underwent HHS's clearance process. Through this process, CDC and NIH, along with HHS's Assistant Secretary for Legislation and its Office of the Assistant Secretary for

Planning and Evaluation, reviewed the documents, and FDA revised the risk assessment and draft guidance to address their comments. CDC, EPA, and NIH officials told us that they were generally satisfied with FDA's coordination efforts and the extent to which FDA addressed their comments. For example, CDC officials said that the agency provided FDA several rounds of comments, and by the end of the process, all of its comments had been considered.

OMB also chose to review FDA's risk assessment and draft guidance on arsenic in infant rice cereal through its interagency review process.[71] According to FDA officials, as part of this process, which occurred from May 2015 through March 2016, FDA coordinated with EPA again, as well as with OMB's Office of Information and Regulatory Affairs and the U.S. Trade Representative within the Executive Office of the President, the Small Business Administration's Office of Advocacy, and USDA. Officials from the Small Business Administration's Office of Advocacy said that they were generally satisfied with the review process and characterized the outcome as typical in that some, but not all, of their suggested changes were accepted.[72]

However, USDA officials raised concerns about FDA involving them too late in the coordination process and about the extent to which FDA addressed their comments. From May 2015 through July 2015, USDA conducted its first review of these documents and provided FDA with comments. USDA had offered to provide FDA with feedback on versions of the risk assessment on several occasions earlier in the process, but FDA did not accept USDA's offers, according to a USDA official. As discussed below, FDA chose to engage USDA later in the process.

In their comments, USDA officials expressed concerns regarding uncertainties and data limitations in the risk assessment and draft guidance on arsenic in infant rice cereal. USDA also raised questions about whether

---

[71] According to OMB staff, OMB's Office of Information and Regulatory Affairs reviewed FDA's risk assessment and draft guidance using its authority to review significant agency guidance documents under Executive Order 12866, as clarified under OMB memorandum M-09-13 (Mar. 4, 2009), and a 2007 bulletin for agency good guidance practices (72 Fed. Reg. 16, pp. 3432-3440 (Jan. 25, 2007)).

[72] OMB staff stated that FDA provided responses to comments and made revisions based on these comments throughout OMB's review.

sufficient data on the link to adverse health effects existed to warrant the draft guidance. Furthermore, USDA stated that because the documents focus solely on rice, instead of addressing risks to the diet as a whole, FDA needs to share clear, consistent, and understandable messages with the public to alleviate fear and misunderstanding related to the risk posed by arsenic in rice. According to USDA officials, FDA did not adequately address their comments in the revised documents, including FDA's communication strategy. However, according to a senior USDA official, in its response to USDA's comments, FDA maintained that, overall, the comments it received from its external peer reviewers—five university researchers—were supportive of the risk assessment and that based on the peer review, FDA did not change its findings or conclusions. According to this USDA official, FDA also noted that there are insufficient data to accurately quantify the risk from arsenic in rice to pregnant women or children but that it decided moving forward with the draft guidance on arsenic in infant rice cereal would be prudent.

FDA and USDA did not agree on USDA's role in developing the risk assessment and the point at which they should begin coordinating on the risk assessment. FDA officials told us that FDA generally considers agencies' expertise in determining whether and when to include them in the development of risk assessments and related documents. FDA did not see USDA as having a role in developing the risk assessment; rather, FDA officials told us that they reached out to USDA after the risk assessment was drafted, when the agency began to consider how to reduce the levels of arsenic in rice during the growing process and the feasibility of industry meeting its draft guidance on arsenic in infant rice cereal. The officials said that FDA met with USDA officials on numerous occasions and invited them to attend additional meetings with various stakeholders. However, according to a senior USDA official, USDA has relevant scientific and technical expertise that should have played a role in developing the risk assessment. According to this official, if FDA had involved USDA earlier in the development process, FDA may have addressed USDA's comments to a greater extent.

We have shown in prior work that agencies can facilitate their collaborative efforts by developing a mechanism for interagency coordination, and a key issue to consider when developing such a mechanism is whether participating agencies have clarified their roles and responsibilities.[73] FDA officials stated that they were not aware of the existence of any mechanism for coordinating risk assessments of contaminants in food, including arsenic in rice, which among other things, could clarify the roles and responsibilities of participating agencies. FDA officials told us that they followed a 2002 report listing guiding principles when developing the risk assessment, but this chapter, which broadly applies to all foodborne contaminants, did not specify the process FDA should follow to coordinate its risk assessment.[74] However, our review of this 2002 report shows that it recommends that FDA encourage active participation and communication with other agencies and stakeholders and collaboration, when appropriate, as part of its risk assessment development process. Although FDA did reach out to USDA, those meetings were after the completion of the risk assessment. By developing a mechanism for working with relevant agencies to identify their roles and responsibilities for coordinating risk assessments of contaminants in food, including arsenic in rice, FDA could have better assurance that it fully utilizes the expertise of all participating federal agencies.

FDA and USDA Coordinated on Developing Methods to Detect Arsenic in Rice to a Limited Extent

FDA and USDA's FGIS and ARS coordinated on the development of detection methods to a limited extent. Officials from FDA and FGIS told us that they began to coordinate in March 2016, when they discovered, in the course of ongoing coordination in another area, that they were each working independently on developing a faster and less expensive detection

---

[73] GAO-12-1022.

[74] Food and Drug Administration, Center for Food Safety and Applied Nutrition, *Initiation and Conduct of All 'Major' Risk Assessments within a Risk Analysis Framework* (2002).

method using the Arsenator. According to FDA officials, FDA became aware of FGIS's interest in developing methods to detect arsenic in rice during a Codex Alimentarius meeting that researchers from both agencies attended. Therefore, the avoidance of potentially duplicative effort occurred as a result of an informal discussion that occurred during this meeting. With regard to ARS, FDA officials told us that FDA did not coordinate with ARS on the development of the hydride generation method but that FDA used its own validated method to provide ARS with actual arsenic concentrations of samples to help ARS test its method. According to ARS officials, ARS did not coordinate with FDA or FGIS when developing its method on hydride generation. According to an FDA official, FDA did not coordinate the development of its current method to detect inorganic arsenic in rice, the faster method for wine and rice, or the laser ablation method with FGIS, ARS, or any other federal agency.

We have shown in prior work that many of the meaningful results that the federal government seeks to achieve, such as those related to protecting food and agriculture, require the coordinated efforts of more than one federal agency.[75] ARS officials told us that from their perspective, there was no reason to coordinate because ARS, FDA, and FGIS are trying to meet different needs with their research. Further, ARS officials told us that coordinating with FDA would blur the distinction between ARS's scientific role and FDA's regulatory role and may imply that ARS has regulatory responsibilities or expertise. However, all three agencies share a crosscutting strategic interest in developing methods for detecting foodborne contaminants, including arsenic in rice. The strategic plans for ARS and FDA's Center for Food Safety and Applied Nutrition include outcomes and strategies related to the development of detection methods for chemical contaminants or residues. Further, FGIS's strategic plan includes a strategy of developing innovative tests to measure grain quality, and according to FGIS officials, they have considered testing inorganic arsenic as part of measuring grain quality. According to FGIS officials, once they began coordinating with FDA on the Arsenator, they saw value

---

[75] GAO-12-1022.

in coordinating and did so for about 9 months before suspending work on the detection method.

We have noted in prior work that interagency mechanisms to coordinate programs that address crosscutting issues may reduce potentially duplicative efforts.[76] However, neither FDA nor USDA has such a mechanism to coordinate the development of methods to detect arsenic in rice or other methods to detect contaminants in food. FDA officials told us that the agency works with USDA research agencies on food safety in an informal manner, and USDA officials told us that they are not aware of any mechanism for coordination and that coordination with FDA generally occurs at the secretarial level because it cuts across a number of USDA agencies. Recently, we also found another example in which FDA and USDA did not coordinate in developing detection methods for other contaminants in foods. FDA and another USDA agency—the Food Safety and Inspection Service—did not coordinate in developing detection methods for drug residues in seafood.[77] By developing a mechanism to coordinate their crosscutting efforts to develop faster and less expensive methods for detecting contaminants in food, including arsenic in rice, FDA and USDA could enhance their ability to use their resources efficiently and avoid engaging in unnecessary and potentially duplicative efforts.

CONCLUSION

NRC and key recent scientific reviews have indicated that long-term ingestion of arsenic may pose a significant risk to human health, and FDA and USDA have taken various actions to manage the risk to human health of arsenic in rice. Their actions are generally consistent with the essential elements we have identified for managing risk, which can help agencies assess threats that could affect the achievement of their goals. For example,

---

[76] GAO-12-1022.

[77] GAO-17-443. In that report, we recommended that FDA and USDA's Food Safety and Inspection Service coordinate and communicate in developing these methods.

both agencies have conducted research on arsenic detection methods, and FDA has issued for public comment a risk assessment on the human health effects from the long-term ingestion of arsenic in rice. In addition, according to FDA officials, because infants are at a higher risk of experiencing some of the health effects of ingesting arsenic, such as neurodevelopmental effects, and the diets of infants are less varied than that of adults, FDA issued a draft guidance regarding arsenic in infant rice cereal. However, FDA officials have not provided a specific timeline for updating the risk assessment in response to newly-available information or for finalizing the draft guidance for infant rice cereal in response to public comments. Both of these documents could help communicate to the public the risk of arsenic in rice, and updating or finalizing them could also help FDA demonstrate its commitment to increasing transparency and accountability by addressing public comments and clarifying its enforcement authority, among other things.

FDA coordinated the development and review of these key documents with several federal agencies, and these agencies were generally satisfied with FDA's coordination efforts. However, USDA raised concerns about being involved too late in the process and the extent to which its comments were addressed. By developing a mechanism for working with relevant agencies to identify their roles and responsibilities for coordinating risk assessments of contaminants in food, including arsenic in rice, FDA could better ensure that it fully utilizes their expertise. Furthermore, FDA and USDA coordinated on the development of arsenic detection methods to a limited extent. Developing a mechanism to coordinate their crosscutting efforts to develop methods to detect contaminants in food, including arsenic in rice, could help FDA and USDA manage their resources and avoid engaging in unnecessary and potentially duplicative efforts.

RECOMMENDATIONS FOR EXECUTIVE ACTION

We are making a total of five recommendations, including four to FDA and one to USDA. Specifically:

- The Commissioner of FDA should develop a timeline for updating the risk assessment on arsenic in rice (Recommendation 1).
- The Commissioner of FDA should develop a timeline for finalizing the draft guidance on arsenic in infant rice cereal (Recommendation 2).
- The Commissioner of FDA should develop a mechanism for working with relevant agencies to identify their roles and responsibilities for coordinating risk assessments of contaminants in food, including arsenic in rice (Recommendation 3).
- The Commissioner of FDA should work with USDA to develop a mechanism to coordinate the development of methods to detect contaminants in food, including arsenic in rice (Recommendation 4).
- The Secretary of Agriculture should work with FDA to develop a mechanism to coordinate the development of methods to detect contaminants in food, including arsenic in rice (Recommendation 5).

AGENCY COMMENTS AND OUR EVALUATION

We provided a draft of this chapter to EPA, HHS, OMB, and USDA for their review and comment. HHS and USDA provided written comments, which are summarized below and reproduced in appendix III and appendix IV, respectively. In addition, EPA, HHS, and USDA provided technical comments, which we incorporated as appropriate. OMB did not comment.

In its comments, HHS generally agreed with our findings and three of the four recommendations directed to it and partially agreed with the other recommendation. Specifically, HHS partially agreed with our first recommendation for FDA to develop a timeline for updating the risk assessment on arsenic in rice, noting that the evolving nature of science precludes it from committing to a specific timeline. We recognize that new scientific studies continue to add to the understanding of the risk of arsenic. However, we continue to believe that FDA should demonstrate its commitment to increasing transparency and accountability by developing a timeline to update the risk assessment, potentially in conjunction with finalizing the draft guidance on arsenic in infant rice cereal. Such an update may state that recent scientific studies or public comments have not resulted in a change to FDA's assessment of the risk.

HHS generally agreed with our findings about the actions it has taken to manage the risk from arsenic in rice and the extent of its coordination with USDA and other agencies. HHS noted that it anticipates developing a final guidance establishing an action level of 100 ppb of inorganic arsenic in infant rice cereal by the end of 2018, which will be consistent with our recommendation. HHS also noted that it will consider ways to enhance mechanisms—such as the Interagency Risk Assessment Consortium—to collaborate and coordinate in the development of risk assessments with agencies that have regulatory responsibility or specific expertise. Further, HHS stated that FDA agrees that a mechanism for better coordinating with USDA on the development of methods to detect contaminants in foods would be worthwhile. FDA will consider whether and how existing mechanisms, such as the lnteragency Residue Control Group and the annual meeting with USDA's ARS and the Food Safety and Inspection Service on food safety research, could be used to improve collaboration with USDA on method development. HHS's plans to enhance or use existing interagency mechanisms may be responsive to our recommendations if they focus on enhancing coordination with other agencies that have expertise or similar goals in the areas of risk assessments and methods to detect foodborne contaminants.

In its comments, USDA generally agreed with our findings and the one recommendation we directed to it. Specifically, USDA generally agreed with our findings about the extent to which FDA coordinated with USDA on the development of methods to detect contaminants in food, including arsenic in rice. It also generally agreed with our recommendation that USDA work with FDA to develop a mechanism to do so and stated that the USDA Office of the Chief Scientist will facilitate this effort. Further,

USDA noted that the Interagency Risk Assessment Consortium may be an appropriate mechanism for addressing GAO's recommendations. USDA's proposal has the potential to be responsive to our recommendation if it focuses on enhancing coordination with FDA regarding the development of detection methods for foodborne contaminants.

As agreed with your office, unless you publicly announce the contents earlier, we plan no further distribution of this chapter until 30 days from the report date. At that time, we will send copies to the appropriate congressional committees; the Secretaries of Agriculture and Health and Human Services; the Administrator of EPA; the Director of OMB; and other interested parties.

Sincerely yours,

Steve D. Morris
Director, Natural Resources and Environment.

APPENDIX I: OBJECTIVES, SCOPE, AND METHODOLOGY

This chapter examines (1) what the National Research Council (NRC) and recent key scientific reviews have reported about the effects of ingestion of arsenic on human health, (2) the extent to which the Food and Drug Administration (FDA) and U.S. Department of Agriculture (USDA) have managed the risk to human health from arsenic in rice, and (3) the extent to which FDA has coordinated with USDA and other federal agencies on actions to manage the risk. In this chapter, we use the term arsenic to refer to either total arsenic or inorganic arsenic. We use the term rice to encompass rice grain and products made with rice, such as infant rice cereal.

To determine what NRC and recent key scientific reviews have reported about the effects of ingestion of arsenic on human health, we analyzed NRC's 2013 report on inorganic arsenic and 14 reviews of the scientific literature published from January 2015 through early June 2017 on the human health effects of ingestion of arsenic.[78] We conducted a literature search of several research databases, such as PubMed and Toxline, to identify reviews that (1) were focused on the effects of ingestion of arsenic on human health; (2) were peer-reviewed; (3) relied on human, rather than animal, studies; (4) provided conclusions or summary statements related to more than one study, rather than just listing individual study findings; (5) included an abstract; and (6) were written in English. We assessed the scientific and statistical credibility, reliability, and methodological soundness of the reviews. We also contacted some of the authors for additional methodological information. Methodological information included, for example, criteria for selecting the studies used in the review; meta-analyses; or meta-regression approach.[79] It also included limitations that the authors cited for the studies they reviewed or for any analyses they conducted. We excluded articles for which we could not clearly determine the methodology. We also reviewed the authors' statements regarding conflicts of interest and determined that none of the articles should be excluded for this reason. We did not examine the references cited by these reviews as part of our analysis. We also did not examine the studies cited by the NRC. The studies we reviewed are listed in appendix II.

To determine the extent to which FDA and USDA have managed the risk to human health from arsenic in rice, we examined relevant provisions in the Federal Food, Drug, and Cosmetic Act, as amended;[80] the Federal Agriculture Improvement and Reform Act of 1996;[81] and other relevant laws, regulations, and policies.

---

[78] National Research Council, Committee on Inorganic Arsenic, *Critical Aspects of EPA's IRIS Assessment of Inorganic Arsenic: Interim Report* (Washington, D.C.: National Academies Press, 2013).

[79] Meta-analyses, including meta-regression, are statistical approaches that combine the results and pool the data from multiple, previous studies on a topic. These techniques allow researchers to draw conclusions and identify patterns that may not be apparent from individual studies.

[80] 21 U.S.C. §§ 301, et seq.

[81] Pub. L. No. 104-127, 110 Stat. 888 (Apr. 4, 1996).

We also used the essential elements for managing risk as identified in our prior work on enterprise risk management.[82] These include: (1) align the risk management process with goals and objectives, (2) identify risks, (3) assess risks, (4) respond to the risks, (5) monitor the risks, and (6) communicate and report on the risks. We identified information on agency actions for managing the risk from arsenic in rice by collecting documentation and interviewing officials from FDA and USDA and we reviewed the information in light of the requirements, policies, and elements. We assessed FDA's and USDA's reported actions to determine the extent to which each agency's actions aligned with these elements. In assessing FDA's and USDA's actions against these essential elements, we used the terms "consistent" and "partially consistent" to reflect the extent to which each agency's actions aligned with an essential element. A determination of "consistent" meant that the agency provided evidence that it had taken major actions in alignment with that essential element. A determination of "partially consistent" meant that the agency provided evidence that it had taken some actions in alignment with that essential element.

We also interviewed 17 stakeholders to obtain their views on the extent to which FDA's and USDA's actions managed the risk, including university researchers (academics) specializing in relevant fields such as epidemiology and soil chemistry, representatives of a consumer organization, and representatives of the rice industry, including rice mills and farms. We identified stakeholders based on suggestions from agency officials and other stakeholders; through our site visit in Arkansas' rice agricultural research and production areas and rice mills; and based on the stakeholders' unique perspective or qualifications, such as membership in the NRC Committee on Inorganic Arsenic. The views we obtained from these interviews are not generalizable to all university researchers or consumer or rice industry organizations but they provide illustrative

---

[82] GAO, *Enterprise Risk Management: Selected Agencies' Experiences Illustrate Good Practices in Managing Risk,* GAO-17-63 (Washington, D.C.: Dec. 1, 2016). This chapter also identified good practices agencies are implementing that illustrate the essential elements of enterprise risk management.

examples of the views of such stakeholders. Table 1 lists information about the 17 stakeholders we interviewed.

Table 1. Information about Selected Stakeholders

Stakeholder	Type of Stakeholder
1. Dr. Aaron Barchowsky, Department of Environmental and Occupational Health, University of Pittsburgh	Academic
2. Dr. Rebecca Fry, Gillings School of Global Public Health, University of North Carolina	Academic
3. Dr. Jarrod Hardke, Division of Agriculture, University of Arkansas	Academic
4. Dr. Brian Jackson, Department of Earth Sciences, Dartmouth Collegea	Academic
5. Dr. Margaret Karagas, Department of Epidemiology, Dartmouth Collegeb	Academic
6. Dr. Keeve Nachman, Bloomberg School of Public Health, Johns Hopkins University	Academic
7. Dr. Ana Navas-Acien, Mailman School of Public Health, Columbia University	Academic
8. Dr. Angelia Seyfferth, Department of Plant and Soil Sciences, University of Delaware	Academic
9. Consumers Union	Consumer organization
10. Organic Trade Association	Rice industry
11. USA Rice Federation	Rice industry
12. Nestlé	Rice industry
13. Lundberg Family Farms	Rice industry
14. Producers Rice Mill, Inc.	Rice industry
15. Riceland Foods, Inc.	Rice industry
16. Vaught Farms, Inc.	Rice industry
17. Brantley Farming Company	Rice industry

Source: GAO analysis of data from selected stakeholders. | GAO-18-199.

Notes:

[a]Dr. Jackson invited Dr. Tracy Punshon of Dartmouth College Department of Biological Sciences to join him in the interview. We incorporated her views as appropriate.

[b]Dr. Karagas invited Dr. Carolyn Murray of the Dartmouth Geisel School of Medicine to join her in the interview. We incorporated her views as appropriate.

To determine the extent to which FDA has coordinated with USDA and other federal agencies on actions to manage the risk to human health from arsenic in rice, we identified relevant actions and examined whether FDA developed interagency collaborative mechanisms, which we have previously reported could help to facilitate coordination between agencies.[83]

To identify actions for which the agencies shared similar goals in their strategic plans or relevant expertise and for which FDA would be expected to coordinate with USDA and other federal agencies, we reviewed relevant provisions in the Federal Food, Drug, and Cosmetic Act, as amended;[84] the Federal Agriculture Improvement and Reform Act of 1996;[85] other relevant laws, regulations, and policies; the current science and research strategic plan for FDA's Center for Food Science and Applied Nutrition and current strategic plans for USDA's Agricultural Research Service (ARS) and Federal Grain Inspection Service (FGIS); and information about the agencies' missions from their websites. These actions were the development of FDA's risk assessment and draft guidance on arsenic in rice and FDA's and USDA's efforts to develop detection methods for arsenic in rice.

We interviewed FDA officials and reviewed documentation they provided to identify the other federal agencies and offices with which FDA coordinated the development and review of its risk assessment and draft guidance on arsenic in infant rice cereal and the development of methods for detecting arsenic in rice. These agencies and offices included ARS, the Centers for Disease Control and Prevention, the Environmental Protection Agency (EPA), FGIS, National Institutes of Health's National Institute of Environmental Health Sciences, the Department of Health and Human Services' Assistant Secretary for Legislation and Office of the Assistant Secretary for Planning and Evaluation, Office of Management and Budget's (OMB) Office of Information and Regulatory Affairs, the Small

---

[83] GAO, *Managing for Results: Key Considerations for Implementing Interagency Collaborative Mechanisms,* GAO-12-1022 (Washington, D.C.: Sept. 27, 2012).

[84] 21 U.S.C. §§ 301, et seq.

[85] Pub. L. No. 104-127, 110 Stat. 888 (Apr. 4, 1996).

Business Administration's Office of Advocacy, and the U.S. Trade Representative.

To determine the extent to which FDA coordinated its risk assessment and draft guidance on arsenic in rice with USDA and other federal agencies, we obtained and reviewed FDA's framework for conducting risk assessments; reviewed agencies' comments on these documents;[86] interviewed FDA officials regarding FDA's efforts to coordinate with other agencies; and interviewed officials from the Centers for Disease Control and Prevention; EPA; the National Institutes of Health; OMB; the Small Business Administration's Office of Advocacy; and USDA regarding the nature of their comments, their experiences coordinating with FDA, and the extent to which FDA addressed their comments.

To examine the extent to which FDA and USDA coordinated the development of arsenic detection methods, we obtained and reviewed documents, including those describing the detection methods that FDA, ARS, and FGIS have developed or have under development, and we interviewed officials from these agencies regarding their efforts to develop these methods and coordinate their development efforts. We also interviewed officials from these agencies to gather their views on the effectiveness of these coordination efforts. We then examined whether FDA had interagency collaborative mechanisms for the development of its risk assessment and draft guidance, and its efforts with USDA to develop arsenic detection methods. We also examined whether participating agencies clarified their roles and responsibilities. Our prior work identified this as a key issue for agencies to consider when implementing coordination mechanisms. We selected this practice because it was relevant to the challenges the agencies faced.

We conducted this performance audit from December 2016 to March 2018 in accordance with generally accepted government auditing standards.

---

[86] USDA and EPA provided us with copies of their comments on FDA's risk assessment and draft guidance on arsenic in infant rice cereal.

Those standards require that we plan and perform the audit to obtain sufficient, appropriate evidence to provide a reasonable basis for our findings and conclusions based on our audit objectives. We believe that the evidence obtained provides a reasonable basis for our findings and conclusions based on our audit objectives.

APPENDIX II: RECENT REVIEWS OF THE HEALTH EFFECTS OF INGESTION OF ARSENIC

The following list identifies recent key reviews of the health effects of ingestion of arsenic that we analyzed.

Amadi, C.N., Z.N. Igweze, and O.E. Orisakwe. "Heavy Metals in Miscarriages and Stillbirths in Developing Nations." *Middle East Fertility Society Journal*, vol. 22, no. 2 (2017): 91-100.

Bardach, A.E., A. Ciapponi, N. Soto, M.R. Chaparro, M. Calderon, A. Briatore, N. Cadoppi, R. Tassara, and M.I. Litter. "Epidemiology of Chronic Disease Related to Arsenic in Argentina: A Systematic Review." *The Science of the Total Environment*, vol. 538, (2015): 802-16.

Karagas, M.R., A. Gossai, B. Pierce, and H. Ahsan. "Drinking Water Arsenic Contamination, Skin Lesions, and Malignancies: A Systematic Review of the Global Evidence." *Current Environmental Health Reports*, vol. 2, no. 1 (2015): 52-68.

Khanjani, N., A. Jafarnejad, and L. Tavakkoli. "Arsenic and Breast Cancer: A Systematic Review of Epidemiologic Studies." *Reviews on Environmental Health* (2017).

Lamm, S. H., H. Ferdosi, E. K. Dissen, J. Li, and J. Ahn. "A Systematic Review and Meta-Regression Analysis of Lung Cancer Risk and Inorganic Arsenic in Drinking Water." *International Journal of Environmental Research and Public Health*, vol. 12, no. 12 (2015): 15498-15515.

Mayer, J. E. and R. H. Goldman. "Arsenic and Skin Cancer in the USA: The Current Evidence regarding Arsenic-Contaminated Drinking Water." *International Journal of Dermatology*, vol. 55, no. 11 (2016): e585-e591.

Milton, A. H., S. Hussain, S. Akter, M. Rahman, T.A. Mouly, and K. Mitchell. "A Review of the Effects of Chronic Arsenic Exposure on Adverse Pregnancy Outcomes." *International Journal of Environmental Research and Public Health*, vol. 14, no. 6 (2017).

Phung, D., D. Connell, S. Rutherford, and C. Chu. "Cardiovascular Risk from Water Arsenic Exposure in Vietnam: Application of Systematic Review and Meta-Regression Analysis in Chemical Health Risk Assessment." *Chemosphere*, vol. 177 (2017): 167-175.

Quansah, R., F. A. Armah, D. K. Essumang, I. Luginaah, E. Clarke, K. Marfoh, S.J. Cobbina, et al. "Association of Arsenic with Adverse Pregnancy Outcomes/Infant Mortality: A Systematic Review and Meta-Analysis." *Environmental Health Perspectives*, vol. 123, no. 5 (2015): 412-21.

Robles-Osorio, M. L., E. Sabath-Silva, and E. Sabath. "Arsenic-Mediated Nephrotoxicity." *Renal Failure*, vol. 37, no. 4 (2015): 542-7.

Sidhu, M. S., K. P. Desai, H. N. Lynch, L. R. Rhomberg, B. D. Beck, and F. J. Venditti. "Mechanisms of Action for Arsenic in Cardiovascular Toxicity and Implications for Risk Assessment." *Toxicology*, vol. 331 (2015): 78-99.

Sung, T., J. Huang, and H. Guo. "Association between Arsenic Exposure and Diabetes: A Meta-Analysis." *BioMed Research International*, (2015).

Tsuji, J. S., M. R. Garry, V. Perez, and E. T. Chang. "Low-Level Arsenic Exposure and Developmental Neurotoxicity in Children: A Systematic Review and Risk Assessment." *Toxicology,* vol. 337, (2015): 91-107.

Von Stackelberg, K., E. Guzy, T. Chu, and B.C. Henn. "Exposure to Mixtures of Metals and Neurodevelopmental Outcomes: A Review." *Risk Analysis*, vol. 35, no. 6 (2015): 971-1016.

APPENDIX III: COMMENTS FROM THE DEPARTMENT OF HEALTH AND HUMAN SERVICES

DEPARTMENT OF HEALTH & HUMAN SERVICES   OFFICE OF THE SECRETARY

Assistant Secretary for Legislation
Washington, DC 20201

MAR 0 1 2018

Steve Morris
Director, Natural Resources and Environment
U.S. Government Accountability Office
441 G Street NW
Washington, DC 20548

Dear Mr. Morris:

Attached are comments on the U.S. Government Accountability Office's (GAO) report entitled, "*Food Safety: Federal Efforts to Manage the Risk of Arsenic in Rice*" (GAO-18-199).

The Department appreciates the opportunity to review this report prior to publication.

Sincerely,

Matthew D. Bassett
Assistant Secretary for Legislation

GENERAL COMMENTS OF THE DEPARTMENT OF HEALTH AND HUMAN SERVICES (HHS) ON THE GOVERNMENT ACCOUNTABILITY OFFICE'S DRAFT REPORT ENTITLED – FOOD SAFETY: FEDERAL EFFORTS TO MANAGE THE RISK OF ARSENIC IN RICE (GAO-18-199)

The U.S. Department of Health and Human Services (HHS) appreciates the opportunity from the Government Accountability Office (GAO) to review and comment on this draft report. The Food and Drug Administration (FDA) takes seriously its mandate to oversee the safety of domestic and imported foods, in part, through the monitoring of contaminants and the assessment of potential exposure and risk, especially for vulnerable populations. We have been testing for arsenic and other contaminants in foods for many years and using data from our monitoring activities to inform the development of interventions to minimize public health risk. As the draft report describes, based on the growing body of scientific evidence on the health risks associated with arsenic exposure and the FDA's testing of rice and rice products, we coordinated with U.S. Department of Agriculture (USDA) and other federal agencies to take steps to minimize the risk to public health. In FY 2016, following our many meetings with manufacturers of infant rice cereal and the publication of our draft guidance, we determined that only two out of 33 samples of infant rice cereal tested had inorganic arsenic levels above the proposed 100 ppb action level. As the draft report mentions, we believe our draft guidance is being used as the industry standard, and we are committed to finalizing it. FDA also is working on a proactive strategy to look broadly at toxic elements in foods, such as arsenic and lead, to be sure that our relatively limited resources are focused on areas where we could have the greatest public health impact.

We value GAO's recognition of the importance of collaborative mechanisms that facilitate coordination between the FDA, USDA, and other federal agencies to manage public health risks such as exposure to arsenic. As evidence of FDA's desire to enhance collaboration, in January, FDA and USDA announced a formal agreement to bolster coordination between the two agencies and aimed at making the oversight of food more efficient and effective in the areas of dual jurisdiction, produce safety, and biotechonology.[1] FDA also relies on ongoing collaborative and coordinating efforts in the areas of research, methods development, and risk assessments across HHS, USDA, and the Environmental Protection Agency. By coordinating and working together with other agencies in these areas, efficiencies can be achieved and each agency's specific food safety and scientific experience can be leveraged to strengthen the federal food safety system.

Recommendation 1

The Commissioner of FDA should develop a timeline for updating the risk assessment on arsenic in rice.

HHS Response

HHS partially concurs with GAO's recommendation.

The FDA is committed to using the best available science to inform and support policy decisions including regulatory actions and consumer messages on public health issues. A risk assessment approach is a useful means of transparently, systematically, and clearly addressing complex risk management questions to inform these decisions. The scope of a specific risk assessment is driven by the specific questions to be addressed. FDA's 2016 risk assessment, "Arsenic in Rice

---

[1] https://www.fda.gov/Food/InternationalInteragencyCoordination/DomesticInteragencyAgreements/ucm594371.htm

and Rice Products Risk Assessment Report," provided the scientific basis to support the regulatory actions and consumer messages. FDA continues to monitor the scientific literature to evaluate whether any new science would change the conclusions of the 2016 risk assessment, requiring FDA to update the risk assessment for the purpose of informing a change in FDA policy or consumer messages. Given the unpredictable, evolving nature of the science, however, we are unable to commit to any specific timeline. For example, and as noted in the draft report, we will continue to collaborate with EPA's Integrated Risk Information System (IRIS) Program on their updated assessment of inorganic arsenic and will consider if EPA's update would necessitate any changes to our risk assessment report correspondent with EPA's findings.

Recommendation 2
The Commissioner of FDA should develop a timeline for finalizing the draft guidance on arsenic in infant rice cereal.

HHS Response
HHS concurs with GAO's recommendation.

The FDA is committed to finalizing the draft guidance establishing an action level of 100 ppb for inorganic arsenic in infant rice cereal. FDA anticipates developing a final guidance by the end of 2018.

Recommendation 3
The Commissioner of FDA should develop a mechanism for working with relevant agencies to identify their roles and responsibilities for coordinating risk assessments of contaminants in food, including arsenic in rice.

HHS Response
HHS concurs with GAO's recommendation.

The FDA will consider ways to enhance its mechanisms to collaborate and coordinate with other agencies in the development of risk assessments for which an agency has a regulatory responsibility and/or specific expertise. FDA has a long history of working collaboratively with other federal agencies with a food safety mission through the Interagency Risk Assessment Consortium. Established by presidential order in 1998, the Consortium is a technical resource for collaboration, efficient use of federal resources, and innovation among member agencies for the development and use of food safety risk assessments and related tools and research. FDA also will seek to enhance awareness of its ongoing risk assessments among other agenices and request they identify their roles and responsibilities with regard to these specific risk assessments.

Recommendation 4
The Commissioner of FDA should work with USDA to develop a mechanism to coordinate the development of methods to detect contaminants in food, including arsenic in rice.

HHS Response
HHS concurs with GAO's recommendation.

The FDA continues to use formal groups/meetings, such as the Interagency Residue Control Group and the annual USDA Agricultural Research Service (ARS) and Food Safety and Inspection Service (FSIS) Food Safety Research meeting to coordinate and develop research projects. Additionally, collaborations among researchers developed through involvement with the national and international scientific community (e.g., Codex, AOAC International, and American Chemical Society) will continue to assist in the development and coordination of research. These informal collaborations were effectively leveraged in the development of both the Arsenator™ and hydride generation collaborative study. Even with these established and successful means of interaction, FDA agrees that a mechanism for better coordinating with USDA on development of methods to detect contaminants in foods would be worthwhile. FDA will consider whether and how these existing collaborations—such as the Interagency Residue Control Group, the annual meeting with USDA's ARS and FSIS on Food Safety Research, other cooperation and coordination groups—could be used to provide improved ways for collaborating with USDA on method development.

APPENDIX IV: COMMENTS FROM THE U.S. DEPARTMENT OF AGRICULTURE

USDA

United States Department of Agriculture

Research Education Economics

Office of the Under Secretary

Room 214W
Jamie L. Whitten Building
Washington, DC 20250-0110

February 16, 2018

Mr. Steve D. Morris
Director, Natural Resources and Environment
United States Government Accountability Office
441 G Street NW.
Washington, D.C. 20548

Dear Mr. Morris:

The United States Department of Agriculture (USDA) appreciates the opportunity to review and provide comments on the draft Government Accountability Office (GAO) report to Congressional requesters GA0-18-199, "FOOD SAFETY: Federal Efforts to Manage the Risk of Arsenic in Rice" (March 2018).

The report was reviewed by the following USDA agencies: Federal Grain Inspection Service (FGIS), Agricultural Research Service (ARS), National Institute of Food and Agriculture (NIFA), Risk Management Agency (RMA), and the Office of the Chief Scientist (OCS). In general, USDA supports the GAO report.

USDA agrees with GAO, that FDA and USDA have taken important actions to manage the risk to human health from arsenic in rice. In particular, USDA ARS scientists and NIFA funded projects: 1) compared commercial rice cultivars for inorganic arsenic (iAs) accumulation in the grain for producers; 2) developed innovative water management practices that lower rice grain iAs levels for producers; 3) genetically mapped the genes responsible for iAs accumulation in rice for breeders; and, 4) developed a more rapid, lower cost method for detecting iAs levels in rice grain for industry.

The report has five recommendations for Executive Action that are addressed to FDA and USDA agencies. Recommendation 5 is specifically directed to USDA. The recommendations are:

1) The Commissioner of FDA should develop a timeline for updating the risk assessment on arsenic in rice.
2) The Commissioner of FDA should develop a timeline for finalizing the draft guidance on arsenic in infant rice cereal.
3) The Commissioner of FDA should develop a mechanism for working with relevant agencies to identify their roles and responsibilities for coordinating risk assessments of contaminants in food, including arsenic in rice.
4) The Commissioner of FDA should work with USDA to develop a mechanism to coordinate the development of methods to detect contaminants in food, including arsenic in rice.
5) The Secretary of Agriculture should work with FDA to develop a mechanism to coordinate the development of methods to detect contaminants in food, including arsenic in rice.

USDA Response to GAO Recommendation 5.

The USDA is committed to strengthening the mechanisms for coordinating the development of methods to detect contaminants in food and associated risks with FDA, including arsenic in rice. Hence, the USDA Office of the Chief Scientist (OCS) will facilitate this effort, and provide the Secretary of Agriculture with annual progress reports on the coordination between USDA and FDA on the mitigation of iAs in rice. These actions support the January 30, 2018, formal Agreement U.S. Secretary of Agriculture Sonny Perdue and FDA Commissioner Scott Gottlieb, M.D. announced at the White House aimed at making the oversight of food more efficient and effective by bolstering coordination between the two agencies. The formal Agreement outlines efforts to increase interagency collaboration, efficiency and effectiveness on produce safety and biotechnology activities, while providing clarity to manufacturers.

For USDA, an important challenge in implementing the GAO recommendations (and the 2018 FDA-USDA agreement) is having a shared operational understanding of coordination. For example, the GAO statement that *"FDA and USDA did not agree on USDA's role in developing the risk assessment and the point at which they should begin coordinating on the risk assessment"* is accurate. Thus, USDA supports Recommendation 5, that USDA work with FDA to develop a mechanism for working together and coordinating research in these areas.

USDA recommends that USDA and FDA partner with the Interagency Risk Assessment Consortium (IRAC) in the coordination of GAO's recommendations. IRAC is a network of U.S. federal agencies with responsibilities and interests in developing food-safety risk-assessment tools or conducting or using food-safety risk assessments. IRAC was mandated by presidential order in 1998 to be a central coordinator and technical resource for collaboration, efficient use of federal resources, and innovation among these agencies.

For its part, the USDA ARS Food Safety Program will continue to hold its annual meeting to which FDA is invited, and participates. USDA ARS will also provide copies of the Food Safety Program annual meeting reports to Office of the Secretary for awareness. Research conducted on detection methods for residues is a major component within this meeting, including arsenic in rice. In addition, all USDA ARS Food Safety Research annual project reports are provided in an extensive document to regulatory agency stakeholders, including FDA, and will be provided to the Office of the Secretary for awareness.

In addition, all USDA ARS and NIFA funded research undergoes external peer review and, where appropriate, requires agreement from regulatory stakeholders. For example, arsenic food safety research that included iAs detection methods development at Beltsville, MD, clearly notes involvement with FDA. Information on all USDA ARS and NIFA approved projects is available on the USDA ARS (https://www.nal.usda.gov/fsrio) and USDA NIFA (https://nifa.usda.gov/) Web sites.

Sincerely,

Chavonda Jacobs-Young

Chavonda Jacobs-Young, Ph.D.
Acting Deputy Under Secretary
Acting Chief Scientist, USDA

In: Food Safety
Editor: Georges Martin
ISBN: 978-1-53616-762-7

Chapter 3

IMPORTED SEAFOOD SAFETY: FDA AND USDA COULD STRENGTHEN EFFORTS TO PREVENT UNSAFE DRUG RESIDUES*

United States Government Accountability Office

ABBREVIATIONS

2008 Farm Bill	Food, Conservation, and Energy Act of 2008
2014 Farm Bill	Agricultural Act of 2014
CBP	U.S. Customs and Border Protection
EU	European Union
FDA	Food and Drug Administration
FFDCA	Federal Food, Drug, and Cosmetic Act
FSIS	Food Safety and Inspection Service
FSMA	FDA Food Safety Modernization Act
HACCP	Hazard Analysis Critical Control Point

* This is an edited, reformatted and augmented version of United States Government Accountability Office; Report to Congressional Requesters, Publication No. GAO-17-443, dated September 2017.

MOU	memorandum of understanding
MRL	maximum residue level
MRM	multi-residue method
NOAA	National Oceanic and Atmospheric Administration
PREDICT	Predictive Risk-based Evaluation for Dynamic Import Compliance Targeting
SRT	Self-Reporting Tool
USDA	U.S. Department of Agriculture

WHY GAO DID THIS STUDY

Most seafood consumed in the United States is imported, and about half of it is raised on fish farms. Because farmed seafood is raised in confined areas and susceptible to infections, farmers may use drugs like antibiotics. The use of unapproved drugs or the misuse of approved drugs may result in unsafe residues in seafood that can cause cancer or allergic reactions, according to FDA, which is charged with ensuring the safety of most seafood. Beginning in April 2016, FSIS became responsible for ensuring the safety of imported catfish.

This chapter examines (1) how FDA helps ensure the safety of imported seafood from unsafe drug residues and ways the agency could strengthen its efforts; (2) how FSIS helps ensure the safety of imported catfish from unsafe drug residues and ways the agency could strengthen its efforts; and (3) the extent to which FDA and FSIS coordinate their oversight efforts. GAO reviewed information from each agency and interviewed agency officials and other key stakeholders.

WHAT GAO RECOMMENDS

GAO is making five recommendations, including that FDA pursue agreements with other countries to test seafood exported to the United States and that FSIS visit a sample of fish farms as part of foreign country on-site audits; and that FDA and FSIS coordinate in developing testing methods and corresponding residue levels for imported seafood. FDA agreed with or partially agreed with two; FSIS partially agreed with two and stated it already addresses a third. GAO disagrees and believes the recommendations should be implemented.

WHAT GAO FOUND

To help ensure the safety of imported seafood from unsafe drug residues, the Food and Drug Administration (FDA) generally depends on the actions of foreign processors and U.S. importers. FDA requires processors and importers to follow its Hazard Analysis and Critical Control Point (HACCP) regulations to identify hazards and the critical control points where the hazards, such as pathogen contamination, are likely to occur and take corrective action. FDA also performs a limited number of (1) inspections of processors and importers each year to ensure HACCP compliance, and (2) tests of imported seafood for contaminants, including unsafe drug residues. FDA could strengthen its efforts to ensure the safety of imported seafood from unsafe drug residues by pursuing agreements with other countries requiring that they test seafood exported to the United States for unsafe drug residues. Under an agency plan, FDA is to coordinate with other countries to increase their capabilities related to the safety of food exported to the United States and better leverage their resources. FDA has used country agreements with respect to pathogen hazards in molluscan shellfish intended for export to the United States. According to FDA officials, it might be worthwhile for the agency to pursue agreements with some countries, but FDA would have to carefully consider a number of factors in determining which countries would be appropriate, which it has not yet done.

In assuming responsibility for inspecting imported catfish, the U.S. Department of Agriculture's (USDA) Food Safety and Inspection Service (FSIS) provided foreign countries and others a transition period—March 1, 2016, through September 1, 2017—before full implementation of its catfish inspection program. Following the transition, countries seeking to continue exporting catfish to the United States are to request equivalence determinations by providing documentation showing that their catfish safety inspection systems are equivalent to the U.S. system. FSIS could strengthen its efforts to ensure the safety of imported catfish. The Agricultural Act of 2014 directs FSIS, in part, to consider the conditions under which catfish are raised, domestically and abroad, but FSIS has not made farm visits a routine part of an equivalence determination. It is not clear how FSIS could consider the conditions under which imported catfish are raised consistent with the act without visiting farms. In addition, during this determination, the agency will already have its inspectors in the foreign country for an on-site audit. FSIS officials generally visit government offices, commercial food processing facilities, and food testing laboratories in a foreign country. Without visiting a sample of farms whose catfish are exported to the United States, FSIS may be missing an opportunity to consider the conditions under which catfish are being raised.

FDA and FSIS took steps to accomplish the transfer of catfish oversight from FDA to FSIS, as called for in the 2014 memorandum of understanding (MOU) that both agencies signed. However, they generally have not coordinated on drug residue testing methods, resulting, in some cases, in differences in drug residue levels used to determine if seafood is unsafe—specifically for unapproved drugs—as called for in the 1984 MOU. Without this coordination, the agencies do not have reasonable assurance that they are consistently protecting consumers from unsafe drug residues.

September 15, 2017

The Honorable Thad Cochran
Chairman
Committee on Appropriations
United States Senate

Dear Mr. Chairman:

The United States has increased the amount of seafood it imports in recent years and is currently the world's second largest importer of seafood, importing from approximately 140 countries. These imports account for over 90 percent of the seafood consumed in the United States, with about half coming from aquaculture (fish farming). In 2015, the United States imported almost 6 billion pounds of seafood, including some of the most popular species with U.S. consumers, such as catfish (about 4 percent of all U.S. seafood imports).[1] Because fish grown in confined aquaculture areas can have high rates of bacterial infections, farmers may treat them with drugs, such as antibiotics and antifungal agents, to increase their survival rates. Once farmers introduce drugs, either in feed or water, drug residues can remain in the fish through harvesting, processing, and consumption. According to the Food and Drug Administration's (FDA) 2008 report to Congress and FDA officials, the residues of some drugs can cause cancer or allergic reactions.[2] In addition, some drugs administered to food-producing animal may cause bacteria of human health concern to become resistant to antibiotics used in humans. As imports of farmed seafood increase, so too do the concerns over the presence of drug residues.[3] The potential use of unapproved drugs, or misuse of approved drugs, adds to these concerns. Effective federal oversight is therefore important to help ensure that imported seafood is safe for U.S. consumers.

Historically, FDA has had the responsibility of ensuring the safety of all seafood consumed in the United States. More broadly, under the Federal Food, Drug, and Cosmetic Act (FFDCA), FDA is generally responsible for ensuring that the nation's food supply is safe, wholesome, sanitary, and properly labeled. All FDA-regulated food products imported

---

[1] Shrimp, salmon, tilapia, and catfish are among the most frequently consumed seafood species in the United States that come primarily from aquaculture. In 2015, the United States imported almost 1.3 billion pounds of shrimp, nearly 0.8 billion pounds of salmon, almost 0.5 billion pounds of tilapia, and over 0.2 billion pounds of catfish.

[2] Food and Drug Administration, *Enhanced Aquaculture and Seafood Inspection—Report to Congress* (Washington, D.C.: Nov. 20, 2008).

[3] Other potential hazards associated with fresh and farmed seafood include pathogens, parasites, decomposition, chemical contaminants, and pesticides. Thorough cooking can eliminate the hazards of pathogens, such as salmonella, and parasites, such as tapeworm.

into the United States must meet the same legal and regulatory requirements as food products produced domestically. The FDA Food Safety Modernization Act (FSMA) provided FDA with additional authorities, such as the authority to suspend the registration of a food facility when there is a reasonable probability of food from the facility causing serious adverse health consequences or death to humans or animals.[4] With some exceptions, food facilities, including seafood facilities, must be registered with FDA to engage in the manufacturing, processing, packing, or holding of seafood for consumption in the United States.

The Food, Conservation, and Energy Act of 2008 (2008 Farm Bill) assigned regulatory responsibility, upon the issuance of final regulations, for the inspection of domestic and imported catfish to the U.S. Department of Agriculture (USDA). Within USDA, the Food Safety and Inspection Service (FSIS) carries out this responsibility. FSIS is also responsible for ensuring that meat, poultry, and processed egg products are safe, wholesome, and accurately labeled. In May 2012, we reported on the potential effects that FSIS's then-proposed catfish inspection program might have on other federal seafood safety inspection programs.[5] We stated that FSIS's catfish inspection program would, in part, further fragment the responsibility for seafood safety and introduce overlap.[6] We also testified on this issue in December 2016.[7] The Agricultural Act of 2014 (2014 Farm

---

[4] Pub. L. No. 111-353, 124 Stat. 3885 (2011). Prior to FSMA, FDA focused on reacting to foodborne illnesses after they occurred. FSMA requires that FDA focus on preventing foodborne illnesses. The law also provides FDA with new enforcement authorities designed to achieve higher rates of compliance with prevention- and risk-based food safety standards and to better respond to and contain problems when they do occur. For example, FSMA required new preventive controls and food safety plans at some food processing facilities and farms, enhanced FDA's capacity to trace foodborne illness outbreaks within the nation's food distribution channels, and expanded FDA's authority to conduct a mandatory recall of contaminated food products.

[5] GAO, *Seafood Safety: Responsibility for Inspecting Catfish Should Not Be Assigned to USDA*, GAO-12-411 (Washington, D.C.: May 10, 2012).

[6] Fragmentation refers to those circumstances in which more than one federal agency is involved in the same broad area of national need and opportunities exist to improve service delivery. Overlap occurs when multiple agencies or programs have similar goals, engage in similar activities or strategies to achieve them, or target similar beneficiaries.

[7] GAO, *Seafood Safety: Status of Issues Related to Catfish Inspection*, GAO-17-289T (Washington, D.C.: Dec. 7, 2016).

Bill), among other things, made all fish of the order *Siluriformes*,[8] hereafter referred to as catfish, subject to FSIS inspection upon the issuance of final regulations. In addition, with the 2014 Farm Bill, Congress reaffirmed its commitment to assigning FSIS the responsibility for inspecting catfish. Prior to FSIS assuming this responsibility, FDA was responsible for ensuring the safety of catfish. In December 2015, FSIS issued a final rule for implementing a catfish inspection program, and in March 2016, FSIS assumed responsibility for inspecting domestic catfish processing facilities. In April 2016, FSIS assumed responsibility for inspecting imported catfish. FSIS considers its catfish program to be in a transitional period until September 1, 2017, at which time FSIS will begin analyzing whether countries exporting catfish to the United States have food safety systems for catfish that provide a similar level of public health protection as found in the United States.

You asked us to examine how FDA and FSIS ensure the safety of imported seafood from unsafe drug residues.[9] This chapter examines (1) how FDA helps ensure the safety of imported seafood from unsafe drug residues and ways the agency could strengthen its efforts, (2) how FSIS helps ensure the safety of imported catfish from unsafe drug residues and ways the agency could strengthen its efforts, and (3) the extent to which FDA and FSIS coordinate their oversight efforts.[10]

To examine how FDA and FSIS help ensure the safety of imported seafood and catfish, respectively, from unsafe drug residues, we examined related legislation and regulations; relevant agency policies, manuals, directives, and guidance; and our past work on federal oversight of seafood safety. We also interviewed agency officials to gain a better understanding of their seafood safety programs. At FDA, we reviewed program data for fiscal years 2010 through 2015, including data on imported seafood

---

[8] The order *Siluriformes* includes the families of *Ictaluridae*, *Clariidae*, and *Pangasiidae*.

[9] Unsafe drug residues may include (1) residues from drugs unapproved for aquaculture use in the United States; (2) residues of approved drugs that exceed allowed levels; or (3) residues of drugs that are approved for use on one seafood species, but are being inappropriately used on another seafood species.

[10] For this chapter, we use “coordination” broadly to include interagency activities that others, and other GAO reports, have variously defined as “cooperation,” “collaboration,” “integration,” or “networking.”

sampling and testing and import refusals. We also reviewed a random sample of 74 foreign processor inspection reports out of a total of 318 reports prepared by FDA for fiscal years 2013 through 2015 to determine whether FDA's processor inspections included visiting farms and laboratories. In addition, we reviewed a random, non- generalizable sample of 9 FDA reports of importer inspections out of a total of 232 reports for fiscal year 2015. At FSIS, we reviewed similar data for catfish from May 1, 2016, when the agency began sampling catfish imports, through July 9, 2017. To assess the reliability of the agencies' data, we, among other things, reviewed controls for the systems that house these data and interviewed agency officials regarding these controls and found these FDA and FSIS data to be sufficiently reliable for the purposes of our reporting objectives. In addition, to gain a better understanding of the agencies' programs, we visited selected U.S. ports of entry in Otay Mesa, California; Long Beach, California; and Newark, New Jersey.[11] The Otay Mesa port is a land port of entry into California and in proximity to the Long Beach port. The New Jersey port is the largest port of entry for seafood products in the United States, and the Long Beach port is the second largest port of entry for seafood in the United States. We also visited an FDA laboratory and district office in Irvine, California, to discuss FDA testing of seafood imports for drug residues with laboratory officials. In addition, we interviewed representatives of the National Fisheries Institute,[12] whose members include more than 200 seafood processors, importers, and exporters. We also interviewed a representative from the Consumer Federation of America,[13] whose members include nearly 300 consumer groups, to obtain their perspectives on the agencies' seafood safety programs. To examine ways, if any, that the agencies could strengthen their efforts, we compared FDA's and FSIS's efforts to help ensure the safety of imported seafood and catfish, respectively, to each other and to

---

[11] We selected the largest and second largest ports of entry for seafood into the United States and selected a land port that is in proximity to one of the sea ports.

[12] According to its website, the National Fisheries Institute is a non-profit organization dedicated to education about seafood safety and other issues.

[13] According to its website, the Consumer Federation of America is an association of non- profit consumer organizations that was established to advance the consumer interest through research, advocacy, and education.

efforts of the European Union (EU) to determine if EU practices for ensuring the safety of imported foods have the potential for enhancing U.S. practices. We discussed the EU's practices with officials from the EU's Food and Veterinary Office (Grange, Ireland) to gain a better understanding of its programs and oversight controls for seafood imports.

To review the extent to which FDA and FSIS coordinate their oversight efforts, we reviewed memorandums of understanding on coordination signed by the agencies and coordination activities between the two agencies and also compared their coordination activities to key practices that can help enhance and sustain federal agencies' collaboration that we previously identified,[14] particularly the practice of identifying and addressing needs by leveraging resources when responsibilities cut across more than one federal agency. We also interviewed FDA and FSIS officials to discuss information for all objectives, including obtaining their views on the opportunities to enhance their programs regarding unsafe drug residues in imported seafood. Appendix I provides additional information on our scope and methodology.

We conducted this performance audit from January 2016 to September 2017 in accordance with generally accepted government auditing standards. Those standards require that we plan and perform the audit to obtain sufficient, appropriate evidence to provide a reasonable basis for our findings and conclusions based on our audit objectives. We believe that the evidence obtained provides a reasonable basis for our findings and conclusions based on our audit objectives.

BACKGROUND

Fishery products (i.e., seafood), including wild catch, aquaculture, and processed fish products, are one of the most highly traded commodities in the world today, and more than half of this commodity originates in developing countries. The EU is the largest market for imported seafood,

---

[14] GAO, *Results-Oriented Government: Practices That Can Help Enhance and Sustain Collaboration among Federal Agencies*, GAO-06-15 (Washington, D.C.: Oct. 21, 2005).

followed by the United States and Japan. Roughly 20 years ago, about 60 percent of the seafood consumed in the United States was imported. As of April 2016, over 90 percent of the seafood consumed in the United States was imported, about half of which was produced from aquaculture operations, also known as fish farms. In 2015, China was the overall leading exporter of seafood to the United States, India was the leading exporter of shrimp, and Vietnam was the leading exporter of catfish. Figure 1 shows the proportion of U.S. imports from the five countries exporting the most seafood to the United States in 2015. The "others" category represents 137 countries that also exported seafood to the United States in 2015.[15] Figure 2 shows the countries that in 2015 were the major exporters of shrimp, salmon, tilapia, and catfish to the United States.

As with seafood more generally, the volume of imported catfish has been increasing in recent years. In 2005, the United States imported over 30 million pounds of catfish. In 2010, the United States imported about 137 million pounds; the major catfish exporters were Vietnam, with 79 percent, and China, with 13 percent. By 2015, total catfish imports were almost 250 million pounds, with Vietnam alone accounting for over 95 percent of all such imports. Domestically, catfish production is concentrated in Alabama, Arkansas, Louisiana, and Mississippi. In general, according to USDA, domestic catfish production has trended downward in recent years, but data in this regard are incomplete. For example, data reported by USDA's National Agricultural Statistics Service showed a general decline in domestic catfish production from 2006 through 2012, including a 36 percent decline from 2010 to 2012, from about 472 million pounds in 2010 to about 300 million pounds in 2012.[16] But this agency did not collect and report similar data for the period 2013 through 2016.

[15] These countries include Australia, Indonesia, and Mexico with seafood exports such as swordfish, shrimp, and snapper.

[16] The domestic volumes are based on "round weight". Round weight is the weight of the whole fish before processing or removal of any part.

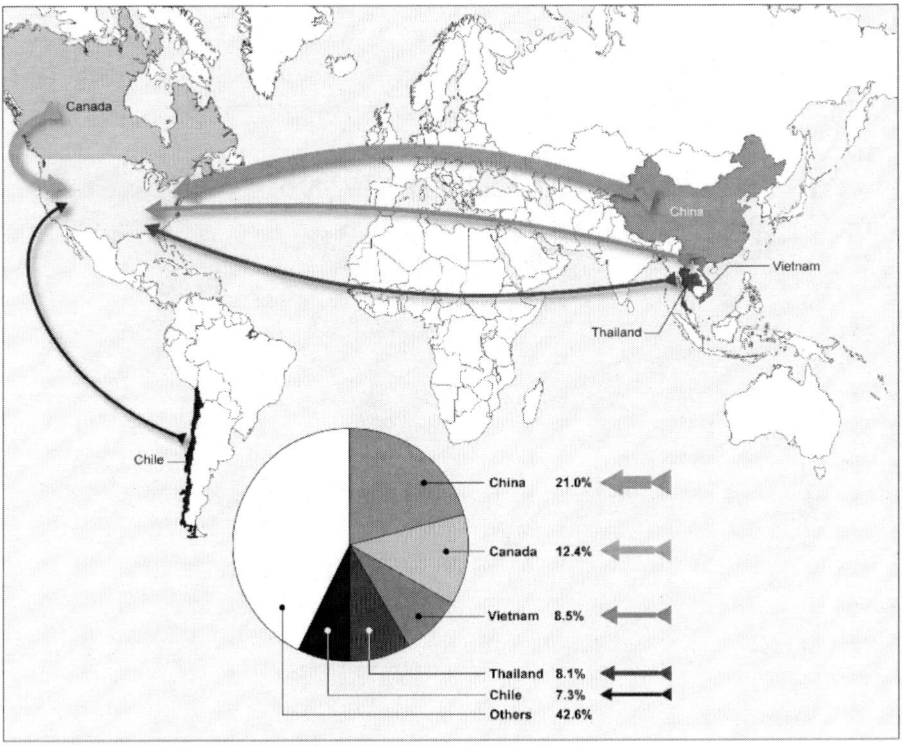

Source: GAO analysis of Department of Commerce and U.S. Customs and Border Protection data; Map Resources (map). | GAO-17-443

Note: The "Others" category represents 137 countries that also exported seafood to the United States in 2015. These countries include Australia, Indonesia, and Mexico with seafood exports such as swordfish, shrimp, and snapper.

Figure 1. Seafood Imports to the United States in 2015 by Country of Origin.

Before 2002, various fish in the order *Siluriformes* were commonly labeled and sold as "catfish." However, in 2002 Congress amended FFDCA to allow only fish from the family *Ictaluridae* (in the order *Siluriformes*) to use the name catfish in labeling. All other fish, such as those from the *Pangasiidae* family (in the order *Siluriformes*) that had previously been labeled as catfish had to have other names on labels, such as basa, swai, or tra. However, as discussed, the 2014 Farm Bill made all fish in the order *Siluriformes* subject to FSIS inspection upon the issuance of final regulations. For purposes of this chapter, we refer to all fish potentially subject to FSIS regulations as "catfish," including fish in the family *Ictaluridae*, which are primarily of domestic origin, and *Pangasiidae*, which come primarily from Vietnam.

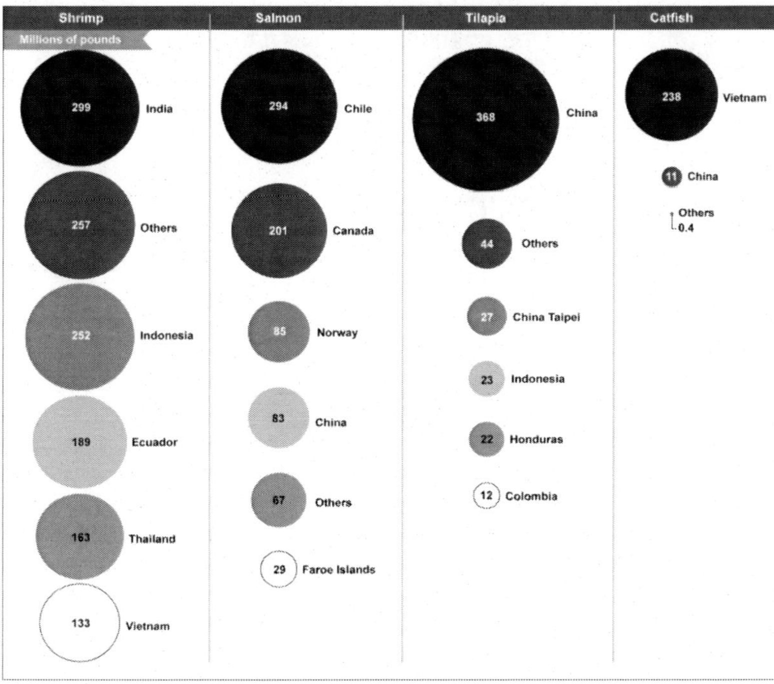

Source: GAO analysis of Department of Commerce and U.S. Customs and Border Protection data. | GAO-17-443

Figure 2. Major Exporters of Shrimp, Salmon, Tilapia, and Catfish to the United States in 2015.

FDA's Approval of Drugs

In the United States, new animal drugs (veterinary drugs) used in animals that are used for food, including seafood, generally must be approved by FDA. According to FDA officials, the process for obtaining a new animal drug approval, including drugs used in aquaculture, originates with an entity or individual (sponsor) submitting an application for review.

Table 1. U.S. Food and Drug Administration (FDA) Approved Aquaculture Drugs and Examples of Their Intended Use in Seafood, as of February 2017

FDA-approved aquaculture drugs	Examples of intended seafood
Anesthetics • Tricaine-S (Tricaine methanesulfonate)	Fish that includes catfish, perch, pike, and salmon
Antibiotics • Aquaflor® (florfenicol) • Terramycin® 200 for fish (oxytetracycline dihydrate) • Oxymarine™, Pennox 343 Soluble Powder, Terramcyin 343®, Tetroxy® 343 Soluble Powder, and Tetroxy® Aquatic (oxytetracycline hydrochloride) • Romet-30® (sulfadimethoxine/ormetoprim) • Sulfamerazine Fish Grade (sulfamerazine)	Freshwater-reared finfish that includes salmon and catfish Includes salmon, catfish and lobster Finfish fry and fingerlings Fish that includes catfish and salmon Rainbow trout, brook trout, and brown trout
Disinfectants • Halamid® Aqua (chloramine-T)	Freshwater-reared fish that includes salmon and trout Walleye Freshwater-reared warmwater finfish that includes catfish and tilapia
Hormones • Chorulon® (chorionic gonadotropin)	Finfish
Parasiticides • Formalin-F, Formacide-B, Parasite-S® (formalin)	Finfish Finfish eggs Penaeid shrimp
Germicidal agents • 35 percent Perox-aid® (hydrogen peroxide)	Freshwater-reared finfish eggs Freshwater-reared fish that includes salmon and trout Freshwater-reared coolwater finfish Channel catfish

Source: GAO analysis of FDA data. I GAO-17-433

Note: Some FDA drug approvals are for specific subsets of these seafood types. For additional information on FDA-approved drugs and their use in seafood, see the following FDA links: http://www.fda.gov/AnimalVeterinary/SafetyHealth/ ProductSafetyInformation/ucm324048.htm and https://animaldrugsatfda.fda.gov/ adafda/views/#/search.

FDA may approve a drug for, among other things, certain disease conditions in specific species (e.g., catfish). When FDA approves a drug, it may establish a tolerance for the safe use of the drug. If residues of the approved drug are detected in a food product above the tolerance, the product is unsafe and therefore adulterated under FFDCA.[17] Once residues are confirmed above a tolerance, which can vary by species and drug, FDA

[17] Adulterated food is, among other things, food that contains a new animal drug (or conversion product thereof) that is unsafe.

may take regulatory action. In addition, if any residue of a drug unapproved by FDA is detected, the food product is likewise unsafe and adulterated under FFDCA. In general, drugs used in aquaculture may include anesthetics, antibiotics, disinfectants, hormones, parasiticides, and germicidal agents. Table 1 lists the drugs that FDA has approved for aquaculture use in the United States as of February 2017.

For testing drug residues, FDA has established three types of maximum residue levels (MRL) for seafood intended for consumption in the United States. FDA applies two of these levels—the "tolerance" and "regulatory" levels—to domestic and imported seafood. The third, "import tolerance" applies only to imported seafood. FDA regards the presence of drug residues above any of these levels to be potentially harmful to human health and may consider taking regulatory action, such as refusing a product's entry into the United States if it contains drug residues above such levels. More specifically, the three levels are as follows:

- Tolerance: For drugs FDA has approved for use in aquaculture, the approval is often for specific species, disease conditions, and a maximum drug residue tolerance. A food product with a residue above the tolerance established for the drug in that food is adulterated under FFDCA. Tolerances for approved drugs are established in regulation and codified in the U.S. Code of Federal Regulations.[18] FDA applies established tolerances for FDA approved drugs to imported and domestic seafood.
- Regulatory level: For certain unapproved drugs, FDA has also established maximum drug residue levels that for purposes of this chapter, we refer to as regulatory levels. In general, these levels correspond to the limits of detection achievable by the testing methods and equipment used by FDA laboratories. FDA does not generally publicly disclose regulatory levels for residues of unapproved drugs. According to agency officials, FDA is concerned that disclosing its regulatory levels for unapproved

---

[18] See Specific Tolerances for Residues of New Animal Drugs, 21 C.F.R. pt. 556, subpt. B.

drugs may encourage some foreign fish farmers to use these drugs if they believe that they can do so in a manner that would result in residue levels below FDA's regulatory levels.

- Import tolerance: For certain drugs that may be used by foreign fish farmers, FDA has established import tolerances. These drugs are not approved for use by domestic fish farmers. FDA has established import tolerances for three drugs[19] that can be used in other countries on seafood, but none of them are for catfish.[20]

The EU's Equivalence Process

The EU, which is the largest importer of seafood worldwide, takes a wide- ranging review of the food safety systems of foreign countries that want to export their seafood products to it. Specifically, the EU requires that other countries seeking to export seafood to it demonstrate that their seafood safety systems meet its or equivalent requirements or comply with specific requirements established in an agreement between the EU and the exporting country. According to EU regulations, for other countries to demonstrate that their seafood safety systems meet EU or equivalent requirements, they must provide information on their food safety systems, including relevant laws, regulatory enforcement powers, and the laboratories that test seafood products. According to EU officials, once they determine that a country has an equivalent food safety system for a particular food product, the government of that country is then responsible for meeting the EU's safety requirements. Moreover, that government becomes the single point of contact to address any identified problems, such as seafood products with drug residues above the EU's accepted MRLs, and is expected to take regulatory actions across that country's supply chain, from farm to processing facility, as necessary.

---

[19] The three drugs for which FDA has developed import tolerances for seafood are lufenuron, azamethiphos, and teflubenzuron.

[20] See http://www.fda.gov/animalveterinary/products/importexports/ucm315830.htm for additional information on import tolerances.

The EU generally conducts an on-site audit of a country's food safety system governing the product that country seeks to export to the EU. The audit focuses on the ability of the competent government authority responsible for food safety to carry out its tasks and provide the necessary guarantees for the safety of food to be exported to the EU. The EU's on-site audits include visiting fish farms and processing facilities as well as reviewing the capabilities and quality of the country's food testing laboratories. In addition, to ensure continuous compliance with EU requirements, the EU periodically conducts follow-up audits of a country's food safety system for specific food products, such as seafood.

Countries that export food to the EU also are required to implement national residue monitoring plans, which include sampling and testing of food products for residues of specific drugs of concern to the EU. The EU must approve a country's residue monitoring plan as a prerequisite for export of food of animal origin, such as seafood, from that country to the EU. Countries exporting food to the EU must submit their residue monitoring plans to the EU for approval each year.[21] The EU can prohibit a country from exporting seafood to the EU if drugs of concern to the EU are not included in the country's national residue monitoring plan or the country does not otherwise fully implement its plan.

FDA Takes Steps to Help Ensure the Safety of Imported Seafood from Unsafe Drug Residues and Could Strengthen Its Efforts with Foreign Country Agreements

To help ensure the safety of imported seafood from unsafe drug residues, FDA generally depends on the actions of foreign processors and U.S. importers. Specifically, FDA requires processors and importers to follow its Hazard Analysis and Critical Control Point (HACCP) regulations to identify hazards and the critical control points where the

[21] Once a country's residue monitoring plan is implemented, the country is to provide an annual report on the sampling results to the EU.

hazards, such as pathogen contamination, are likely to occur and to take corrective action. FDA also performs a limited number of (1) inspections of processors and importers each year to ensure HACCP compliance, and (2) samples and tests of imported seafood for contaminants, including unsafe drug residues. However, FDA has not entered into agreements with other countries requiring that they test seafood exported to the United States for the specific drugs the agency tests for—drugs of concern to FDA.

FDA Requires Seafood Processors and Importers to Follow HACCP and Conducts a Limited Number of Inspections to Assess Compliance

Since 1997, FDA has used HACCP as its main safety oversight tool for imported seafood. Under HACCP requirements, seafood processing firms, including firms that manufacture, pack, or label, are responsible for conducting a hazard analysis and for developing and implementing HACCP plans whenever an analysis shows that one or more hazards, such as pathogens or chemicals, are reasonably likely to occur.[22] Under HACCP requirements, processors are also responsible for addressing hazards that may have been introduced before the seafood reached their facilities, including hazards from unsafe drug residues. FDA inspects a limited number of foreign seafood processing facilities each year to (1) identify potential seafood safety problems before seafood products arrive in the United States, (2) help the agency make admissibility decisions when seafood products are offered for importation into the United States, and (3)

---

[22] A HACCP plan is a written plan that defines the procedures for maintaining control of potential food safety hazards at the critical control points of food preparation or processing. It includes information on the potential hazard associated with a specific food product; the measure that will be implemented to control the hazard; the critical control point to implement the measure; minimum or maximum values (critical control limit) at which a physical, chemical, or biological parameter must be controlled to minimize the risk that a potential food safety hazard may occur; the monitoring procedures to ensure that the hazard is controlled; and the corrective actions to be taken in response to deviations from critical control limits.

help ensure that imported seafood products under FDA's jurisdiction meet U.S. safety requirements, among other things.[23]

In fiscal year 2016, FDA inspected 144 foreign seafood processer facilities for HACCP compliance, or nearly 2 percent of the 7,669 total processors that exported seafood to the United States that year.[24] According to FDA officials, limited resources dictate the number of inspections the agency can conduct. Similarly, in our January 2015 report, we found that FDA was not keeping pace with FSMA's mandate for increasing the number of these inspections of foreign food facilities.[25] According to FDA officials, the agency did not plan to meet this FSMA mandate because of limited funding. Regarding potential drug residues, FDA's inspections of foreign seafood processor facilities are limited. These inspections involve reviewing the processors' HACCP plans and other records to ensure that the processors considered drug residues as a hazard that is reasonably likely to occur if the seafood products it receives are from fish farms. However, the inspections do not include visiting the fish farms that supply the processor facilities to evaluate drug use or controls; it is at these farms where drugs are introduced. FDA's inspections also do not include visiting laboratories that may be asked by processors to test for unsafe drug residues in seafood to assess the laboratories capabilities and competence.

According to FDA officials, neither fish farms nor laboratories are subject to the HACCP regulations and therefore are not part of FDA's processor inspections. Our review of a random sample of 74 foreign processor inspection reports prepared by FDA for fiscal years 2013 through 2015 generally confirmed that FDA's processor inspections did

---

[23] According to FDA documents, inspections should focus on the implementation of the HACCP program for those targeted products. It also includes a review of monitoring, corrective action, and sanitation monitoring records. FDA may also conduct inspections to verify corrective actions have been implemented.

[24] According to FDA officials, the agency only tracks the most recent number of registered processors and does not track the numbers that were registered in the past.

[25] GAO, *Food Safety: Additional Actions Needed to Help FDA's Foreign Offices Ensure Safety of Imported Food*, GAO-15-183 (Washington, D.C.: Jan. 30, 2015). We recommended that FDA complete an analysis to determine the annual number of foreign food inspections that is sufficient to ensure comparable safety of imported and domestic food. If the inspection numbers from that evaluation are different from the inspection targets mandated in FSMA, FDA should report the results to Congress and recommend appropriate legislative changes. FDA concurred with the recommendation.

not include visiting farms that supply the fish to the processor and laboratories that conduct testing for the processor. From the 74 reports, we found only one case where the FDA inspectors also visited a fish farm supplying the processor. In that case, the farm was located near the processing facility.

FDA also requires U.S. importers to verify that foreign processing facilities are HACCP compliant. HACCP regulations require all importers of seafood to the United States to demonstrate that the seafood they import has been processed in accordance with HACCP requirements. For example, under HACCP requirements, importers of seafood products must (1) obtain seafood products from processing facilities in foreign countries that have an agreement with FDA documenting that the country's food safety system complies with or is equivalent to the U.S. system; or (2) maintain written verification procedures that include product specifications designed to ensure that products are not adulterated and take at least one of six affirmative steps to document that the foreign processing facilities supplying their seafood products comply with HACCP requirements. These steps may include, for example, maintaining on file a copy of the foreign processor's HACCP plan and a written guarantee from the processor that the imported seafood is processed in accordance with HACCP requirements, or obtaining a certificate from a foreign government inspection authority or competent third party certifying that the imported fish was processed in accordance with HACCP requirements.

As of June 2017, FDA had no equivalence agreements with other countries, according to FDA officials, precluding importers of seafood products from using the first of the two options noted above. Instead, according to FDA officials, U.S. seafood importers generally comply with HACCP regulations by obtaining (1) a copy of a foreign processor's HACCP plan and an attestation that the firm processes its seafood products in compliance with HACCP regulations and (2) lot-by-lot certifications from either a foreign government authority or third-party auditor attesting to the foreign processor's compliance with HACCP regulations.[26]

---

[26] According to an FDA document, how a lot is defined is largely up to the processor and the importer. For import shipments, FDA characterizes a lot as an entry, group of entries, or a

However, according to FDA officials, under HACCP regulations, importers are not required to either visit foreign seafood processors to ensure that these processors implement their HACCP plans or assess the competence or ability of foreign governments or third- party auditors to issue lot-by-lot certifications.

Each year FDA inspects a small number of U.S. seafood importers to determine if they have obtained the appropriate documentation from foreign seafood processors indicating that the processors had met HACCP requirements. For fiscal years 2010 through 2015, FDA inspected, on average, 178 seafood importers annually, or about 4 percent of the average number of seafood importers (4,009) registered with FDA each year during that period. As with processing facility inspections, FDA officials stated that the number of importer inspections was affected by limited resources. Our review of nine random—a non- generalizable sample—FDA reports of importer inspections for fiscal year 2015 showed various problems with importer compliance with HACCP requirements. For example, of the nine reports, five noted that the importer inspected did not identify the affirmative step taken to document that the foreign processors supplying some or all of that importer's seafood products were in compliance with HACCP requirements. In some cases, the inspection report also noted that an earlier FDA inspection had also found that the importer had failed to identify an affirmative step that was required in order to comply with HACCP regulations.

FDA Conducts a Limited Number of Port-of-Entry Examinations and Tests a Small Sample of Seafood Imports to Identify Unsafe Drug Residues

FDA uses a computerized tool for admissibility screening and to determine, in part, which seafood products to examine.[27] This tool is called

---

portion of an entry of merchandise that can be clearly defined as a shipment for FDA sampling and examination purposes.

[27] Admissibility screening is intended to ensure that the food is admissible under FFDCA. Imported food products are generally considered admissible if they are in compliance with

the Predictive Risk-based Evaluation for Dynamic Import Compliance Targeting (PREDICT).[28] We previously reported on FDA's use of PREDICT in May 2016.[29] Examinations may include physical inspection (e.g., appearance and smell) of the seafood; inspection of the label; or sampling and laboratory testing of the seafood to identify any contaminants, including unsafe drug residues, that may render the seafood adulterated. FDA tests imported seafood for all drugs of concern to the agency and compares results to its corresponding MRL if one exists, according to FDA officials. FDA uses a multi-residue method (MRM) that tests for 26 drugs. FDA also uses other methods to test for drug residues. In cases in which the MRL is exceeded, the imported seafood is adulterated under FFDCA and subject to refusal.

If FDA finds adulterated seafood through testing or examination, the agency can refuse the shipment's entry into the United States. From fiscal years 2010 through 2015, FDA refused entry (import refusal) into the United States of 1,726 seafood products for drug-related violations. The majority of the refusals were of exports from four countries: China (37 percent), Malaysia (28 percent), Indonesia (12 percent), and Vietnam (11 percent). Shrimp represented 54 percent of all FDA refusals; other seafood, 32 percent; tilapia, 7 percent; catfish, 6 percent; and salmon, less than 1 percent. FDA officials said that import refusals may also occur based on FDA inspections of foreign processors but that they were not aware of a refusal based on FDA inspectional findings related to the use of an unsafe drug. Appendix II provides our analysis of FDA's drug testing results for fiscal years 2010 through 2015.

---

applicable FFDCA regulations that ensure food is not adulterated, misbranded, manufactured or packed under insanitary conditions, or restricted in sale in the country in which it was produced or from which it was exported, among other things.

[28] PREDICT is designed to estimate the risk of imported products using information such as the history of the associated foreign processor facility; results of any FDA inspections of the facility, including any past problems with products from the facility; and the country of origin.

[29] GAO, *Imported Food Safety: FDA's Targeting Tool Has Enhanced Screening, but Further Improvements Are Possible*, GAO-16-399 (Washington, D.C.: May 26, 2016). We recommended that FDA take actions to improve the effectiveness of PREDICT by documenting the process by which FDA is to identify, obtain, and use open source data and establishing a timeline for implementing the remaining recommendations from FDA's 2013 evaluation of PREDICT. FDA generally agreed with these recommendations.

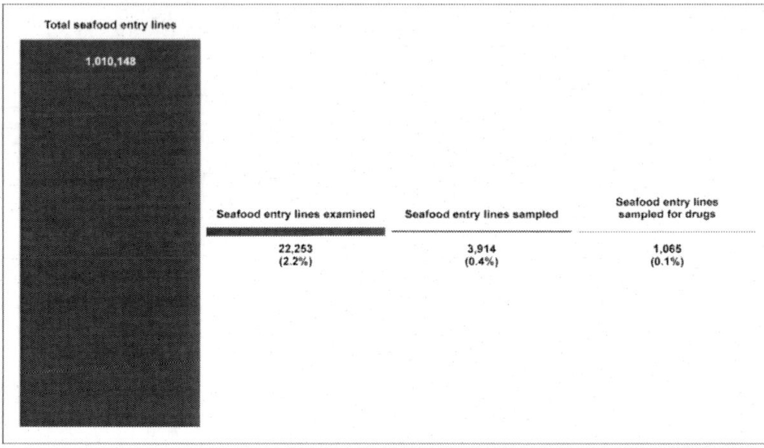

Source: GAO analysis of FDA data. | GAO-17-443

Notes: An entry line is a portion of an import shipment that is listed as a separate item on an entry document. An entry may include one imported item or hundreds; each item is identified as an entry line. Examined entry lines include those subject to physical inspection, label review, and sampling. FDA's selection of the seafood entry lines it sampled was not based solely on drug residue risk concerns but on a combination of risk variables, including potential pathogen hazards. The percentages are of the total seafood entry lines.

Figure 3. Total Seafood Entry Lines, Lines Examined, Lines Sampled, and Lines Sampled for Drugs by the Food and Drug Administration (FDA), Fiscal Year 2015.

In fiscal year 2015, FDA examined 2.2 percent of all imported seafood entry lines for a variety of food safety issues.[30] The number of examinations is limited by available resources, according to FDA officials. Further, regarding drug residues, in fiscal year 2015 FDA tested 0.1 percent of about 1 million seafood entry lines for drugs of concern to FDA in an effort to detect unsafe residues. Of the imported seafood tested, the percentage that tested positive, by type of seafood was as follows: catfish, about 9 percent (3 of 33 samples); salmon, 0 (0 of 86); shrimp, about 12 percent (67 of 550); tilapia, about 11 percent (28 of 258); and "other" seafood, about 7 percent (14 of 213).[31] Based on this level of testing,

---

[30] An entry line is each portion of an import shipment that is listed as a separate item on an entry document offered for admission into U.S. commerce.

[31] The number of samples that FDA tested and the associated results are not intended to be statistically valid for projection purposes. Instead, FDA uses PREDICT for admissibility

seafood shipments from a foreign processing facility would have a roughly 1 in 1,000 chance of being selected by FDA for drug residue testing, unless other information in PREDICT, such as the results of sampling of prior shipments, processing facility inspection records, and country of origin, among other factors, elevated or lowered that facility's risk score. According to FDA officials, the level of testing of seafood imports for drug residues is affected by the availability of resources. Figure 3 shows fiscal year 2015 data for FDA's examinations and sampling of imported seafood, including sampling specifically for drug residues.

Regarding imported catfish specifically, FDA's sampling and testing for unsafe drug residues generally increased from fiscal years 2010 through 2012, but then dropped significantly in the following 3 fiscal years, even as the volume of catfish imports remained relatively the same. For example, as shown in Figure 4, from fiscal years 2012 through 2015, FDA's sampling and testing of imported catfish for unsafe drug residues declined by 75 percent, even as the volume of catfish imports averaged about 250 million pounds annually during this period.

Further, in comparison to FDA's sampling and testing of other popular imported seafood species, such as salmon, shrimp, and tilapia, for unsafe drug residues, FDA's sampling and testing of imported catfish showed the greatest drop relative to import volume for fiscal years 2010 through 2015. For example, as shown in Figure 5, FDA's sampling and testing of imported catfish experienced the greatest drop per 10 million pounds of imports, particularly from fiscal years 2012 through 2015. However, as shown in the figure, the level of sampling and testing per 10 million pounds of imports is limited for all of the seafood species depicted. Appendix II provides information on the total number of imported seafood samples, including for catfish, FDA tested for fiscal years 2010 through 2015.

---

screening and for determining which products to examine and test based on various factors, such as past violations history and country of origin. In addition, some of these sample sizes are particularly small and the percentages may convey a level of precision that can be misleading because they can greatly change with minor changes in the data. The "other" seafood category includes, for example, crab, frog legs, and lobster.

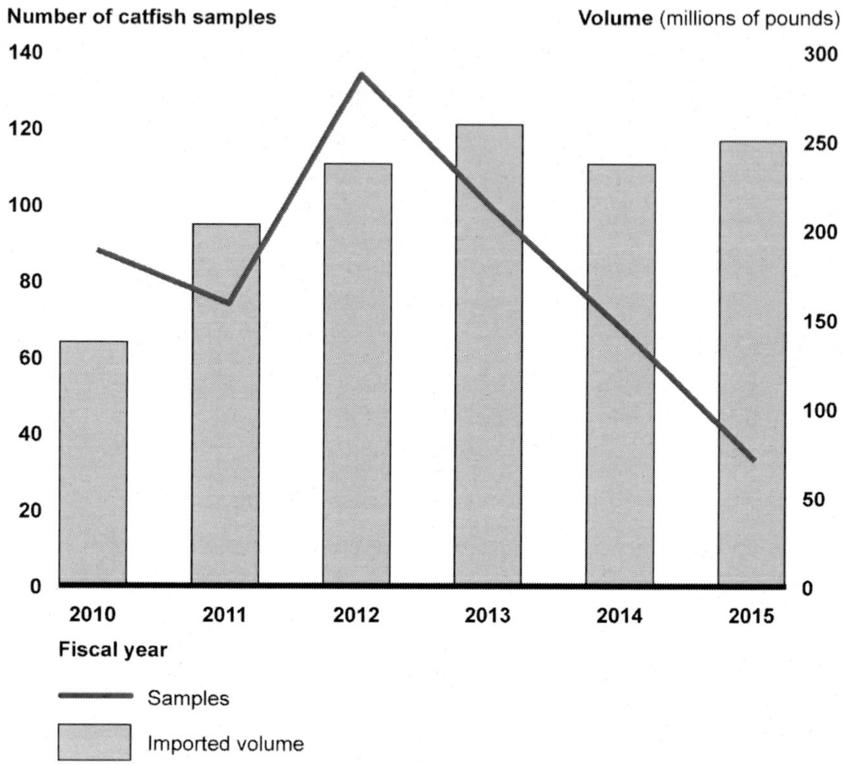

Source: GAO analysis of FDA, Department of Commerce, and U.S. Customs and Border Protection data. | GAO-17-443

Figure 4. Number of Imported Catfish Samples Tested by the Food and Drug Administration (FDA) for Unsafe Drug Residues Compared to the Volume of Catfish Imported, Fiscal Years 2010 through 2015.

The agency decreased its sampling and testing of imported catfish from fiscal years 2012 through 2015, as shown in Figures 4 and 5. According to FDA officials, the agency reduced sampling and testing for fiscal years 2014 and 2015 because it anticipated that FSIS would soon assume this responsibility. As discussed, the 2008 Farm Bill assigned USDA this responsibility upon the issuance of regulations, and the 2014 Farm Bill reaffirmed this decision and gave USDA a timeline for implementing its catfish inspection program (within USDA, this responsibility is delegated to FSIS). FSIS began inspecting domestic catfish in March 2016 and imported catfish in April 2016. FDA officials said that the agency's ability to increase its sampling and testing of imported seafood for unsafe drug residues is constrained by resource limitations and competing priorities in other areas. For example, an E. coli outbreak in this country would require FDA to address that outbreak and potentially reduce resources for drug residue testing.

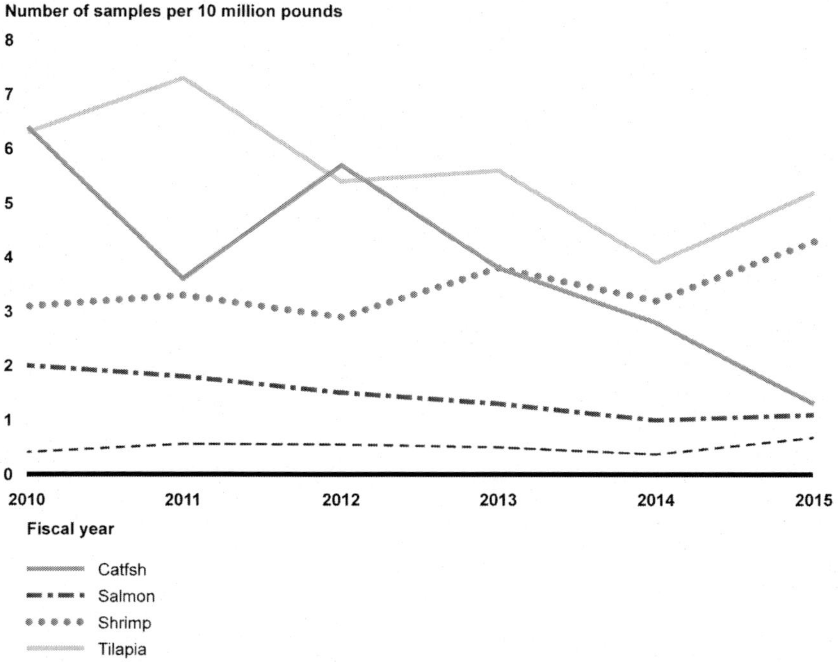

Source: GAO analysis of FDA, Department of Commerce, and U.S. Customs and Border Protection data. | GAO-17-443

Note: Other includes, for example, crab, frog legs, and lobster.

Figure 5. Number of Imported Seafood Samples, by Type, Tested for Drug Residues by the Food and Drug Administration (FDA), per 10 Million Pounds of Imports, Fiscal Years 2010 through 2015.

FDA Could Strengthen Its Efforts by Pursuing Agreements with Other Countries Requiring That They Test Seafood Exported to the United States

FDA is not pursuing agreements with other countries requiring that they test seafood exported to the United States for unsafe drug residues, according to FDA officials, as the EU does. Establishing such agreements is a practice that we have previously recommended to FDA.[32] FSMA requires FDA to develop a comprehensive plan to expand the technical, scientific, and regulatory capacity of foreign governments and their respective food industries in countries that export foods to the United States. To meet FSMA's requirement, in 2013 FDA developed its International Food Safety Capacity-Building Plan. This plan states, in part, that FDA will (1) coordinate with other countries to, among other things, increase their capabilities related to the safety of the food they export to the United States and (2) better leverage other countries' resources. Such leveraging may enable FDA to perform its work to ensure food safety more efficiently. Further, the plan states that agreements and other arrangements with foreign regulatory authorities or other entities are useful in ensuring the safety of food products and avoiding duplication of oversight efforts.

Even with the development of this plan in 2013, as of June 2017 FDA had not entered into any agreements or other arrangements to involve another country in ensuring the safety of the seafood it exports to the United States from unsafe drug residues, according to FDA officials. Further, as of that time, FDA was not actively pursuing such agreements or other arrangements. When asked why, agency officials said that it might be worthwhile for FDA to pursue agreements with some countries, but the

[32] In January 2001, to better ensure the safety of imported seafood, we recommended that FDA develop specific goals and time frames for establishing agreements with other countries to document that their seafood safety systems are equivalent to that of the United States. GAO, *Food Safety: Federal Oversight of Seafood Does Not Sufficiently Protect Consumers*, GAO-01-204 (Washington, D.C.: Jan. 31, 2001). FDA disagreed that it should make it a priority to establish equivalence or other agreements. In a January 2004 report, we said that we continued to believe that such agreements are one of the most cost-effective methods for ensuring the safety of imported seafood. See GAO, *Food Safety: FDA's Imported Seafood Safety Program Shows Some Progress, but Further Improvements Are Needed*, GAO-04-246 (Washington, D.C.: Jan. 30, 2004).

agency would have to carefully consider a number of factors in determining which countries would be appropriate and has not done so. According to these FDA officials, some countries exporting seafood to the United States might not be ready for such agreements. For example, they said that these countries may lack the laboratory infrastructure and capabilities needed to test for the drugs of concern to FDA, including at the corresponding MRLs that the agency established for these drugs, and their laboratory staff may need additional training and education.

According to EU officials, the foreign laboratories that test seafood products exported to the EU for drug residues should ideally be accredited and have the required methods included in the scope of accreditation. Further, these officials stated that those methods must be validated or the residue monitoring plan will not be approved by the EU. Countries seeking to export seafood to the EU must either demonstrate that their seafood safety systems meet EU or equivalent requirements or comply with specific requirements established in an agreement between the EU and the exporting country. The United States may still be a market for at least some seafood that did not meet the seafood safety standards of the EU because FDA does not have agreements with these countries.[33]

Moreover, precedent exists for FDA's use of country agreements in another area. Specifically, the agency has used them with respect to pathogen hazards in molluscan shellfish intended for export to the United States. For example, FDA entered into an agreement with Mexico in 2012 in which Mexico committed that its sanitation program guidelines for harvesting, processing, transporting, and labeling molluscan shellfish would comply with U.S. requirements to reduce potential pathogen hazards. In addition, under the agreement, Mexico committed that its competent government authority would restrict the harvest of molluscan shellfish from unapproved growing areas and take enforcement action against persons or companies harvesting from unapproved areas.

---

[33] In our 2011 report (GAO-11-286), we found that according to an FDA review of Vietnamese documents, all seafood processing facilities' HACCP plans stated that if a drug unapproved by the EU was found in a seafood product, that product should be diverted to another market. FDA officials concluded that this HACCP plan requirement could result in such products being exported to the United States.

In addition, in the course of doing work for our September 2012 report, officials from three major seafood exporting countries described the benefits of countries entering into agreements with them.[34] Noting that they had such agreements with the EU, these officials said that their countries would be open to establishing similar agreements with the United States. For example, see the following:

- Thai officials said that because Thailand has a formal (government-to- government) agreement with the EU, the Thai government can better ensure that its seafood processing facilities meet the EU's safety standards for seafood exported to EU countries. These officials also said that, because of this agreement, their government can more efficiently address problems with Thai seafood exports identified by an EU country, including any drug residue problems.
- Noting that their country has a formal agreement with the EU, Ecuadoran officials said staff from their government inspect and certify processor facilities that export seafood products to the EU. Further, these officials said the Ecuadorian government developed a national control plan to address specific EU requirements and standards for seafood exports, including those related to drug residues.
- Noting that Indonesia has formal agreements with other countries, including the EU, Indonesian officials said that all Indonesian processing facilities that export seafood to those countries must meet HACCP certification requirements; obtain a health certificate; and meet any additional requirements of the importing country, which may include requirements related to drug residues.

Further, the results of FDA's "country assessments" of other countries exporting seafood to the United States point to the potential benefit of having formal agreements with these countries. FDA conducts these

[34] GAO, *Food Safety: FDA Can Better Oversee Food Imports by Assessing and Leveraging Other Countries' Oversight Resources,* GAO-12-933 (Washington, D.C.: Sept. 28, 2012).

assessments to provide the agency with a broad view of a country's regulatory infrastructure and the capacity of its seafood industry to control unsafe drug residues. According to FDA guidance,[35] when the agency determines there is a significant risk that a country may be exporting seafood with unapproved drug residues to the United States, FDA may undertake an assessment of the country's drug residue control program. FDA has conducted 10 foreign country assessments focused on the use of drugs in aquaculture in eight countries from 2006 through December 2016. The most recent assessments of China and Indonesia, major exporters of seafood, were conducted in 2006 and 2007, respectively. According to the guidance, FDA's determination may be based on the results of its own sampling and testing of seafood exported from that country or on other information, such as a significant change in a country's competent government authority responsible for seafood safety oversight. As described in the guidance, in the course of doing an assessment, FDA officials may visit seafood processors, fish farms, drug testing laboratories, feed mills, and veterinary drug distributors and retailers in the country.

At the conclusion of a country assessment, FDA may offer suggestions or recommendations to government officials and seafood industry representatives for improving the country's seafood safety program. However, as noted in some of these assessments, the country involved reported that it could not act on these suggestions or recommendations in the absence of a formal agreement with the United States, or it was otherwise apparent that the absence of an agreement was likely an impediment. For example, see the following:

- In its 2012 assessment of Vietnam, FDA recommended that the Vietnamese government reinstate a requirement that its processors test all farmed seafood consignments intended for the U.S. market for unsafe drug residues until such testing showed that unsafe drug residues were no longer a problem. However, according to the assessment, Vietnam responded that under its new food safety law,

[35] The guidance is found in FDA's *Standard Operating Procedures for Conducting Country Assessments for Animal Drug Residues in Aquacultured Products.*

it can only conduct mandatory consignment testing when required by a formal agreement with the receiving country. Vietnam noted that it has such agreements with other countries, such as Canada, the EU, Japan, and Korea, but does not have an agreement with the United States.

- In its 2013 assessment of Ecuador, FDA strongly suggests that the competent government authority require that seafood processors exporting to the United States meet all requirements in Ecuador's drug residue plan. However, according to the assessment, Ecuador responded that its plan is voluntary and thus the country cannot require that its processors comply with the plan's requirements except in cases where Ecuador has a formal agreement with another country. Ecuador noted that it has such agreements with Argentina, Brazil, China, the EU, and Russia but not with the United States. Those countries that have agreements with Ecuador are regulated by EU standards, according to the assessment. As a result, shrimp or tilapia that cannot be shipped to EU countries because of concerns about violative drug residues may be shipped to the United States.
- In its 2011 assessment of Malaysia, FDA said that Malaysian testing of seafood exported to the United States was limited to three drugs and it was not possible to verify that this seafood met U.S. requirements and was free of unsafe drug residues. In response to FDA's concerns, the competent government authority declined to make changes in its testing, stating it is the responsibility of U.S. importers to ensure that seafood imports comply with U.S. requirements. In contrast, FDA determined that the Malaysian seafood safety control system underwent major changes in response to an EU audit, and the country is on the list of countries approved to export fishery products to the EU.
- In its 2010 assessment of India, FDA noted that drug residue testing required by the Indian government for seafood exports to the EU and the United States differed. Specifically, a greater level of testing was required for seafood exports to the EU. According to

> FDA's assessment, the Indian government said the type of testing done on seafood intended for the EU would also be done on seafood intended for the U.S. market if FDA required that seafood exports be accompanied by a health certificate, as is required by the EU, Japan, and South Korea. However, FDA does not require health certificates and took no action to require them after completing this assessment, according to FDA officials.

As of June 2017, FDA had not entered into any agreements or other arrangements with any country for ensuring the safety of the seafood exported to the United States from unsafe drug residues.

In light of FDA's limited inspections of foreign seafood processors and U.S. importers, its limited sampling and testing of imported seafood for unsafe drug residues, and the limitations of FDA's country assessments to obtain other countries' voluntary cooperation in ensuring that their seafood exports to the United States are free of unsafe drug residues, FDA not pursuing formal country agreements is counter to the agency's International Food Safety Capacity-Building Plan. Under the plan, FDA is to coordinate with other countries to, among other things, increase their capabilities related to the safety of food exported to the United States and better leverage other countries' resources. Such leveraging may enable FDA to more efficiently perform its work to ensure food safety. However, FDA has not pursued formal agreements with countries exporting seafood to the United States that commit the countries to take actions, such as testing for drugs of concern to FDA and the corresponding MRLs, that would leverage those countries' resources, according to FDA officials. The EU and other countries have successfully pursued such agreements. By pursuing formal agreements with countries exporting seafood to the United States that commit these countries to test for drugs of concern to FDA and the corresponding MRLs that the agency established for these drugs, FDA could strengthen its efforts toward ensuring that imported seafood does not contain unsafe drug residues. Further, such agreements would allow FDA to leverage other countries' resources to improve imported seafood safety and further protect U.S. consumers from unsafe drug residues. FDA officials acknowledged that such agreements would be helpful in protecting U.S. consumers from unsafe drug residues.

FSIS Takes Steps to Help Ensure the Safety of Imported Catfish from Unsafe Drug Residues and Could Strengthen Its Efforts

In assuming responsibility for inspecting imported catfish, FSIS provided for an 18-month transition period—March 1, 2016, through September 1, 2017—to provide foreign countries, importers, and other stakeholders time to transition to the full implementation of the agency's catfish inspection program for imports.[36] Following the transition period, a foreign country seeking to continue exporting catfish to the United States after September 1, 2017, is to request an equivalence determination and provide documentation showing that its catfish safety inspection system is equivalent to that of the United States. However, FSIS has not made farm visits a routine part of initial equivalence determination and verification on-site audits. FSIS also does not plan to require other countries to test catfish exported to the United States for the same drugs it tests for—the drugs of concern to FSIS—at the agency's MRLs.

Upon Assuming Responsibility for Catfish Oversight, FSIS Allowed an 18-Month Transition Period

As of March, 1, 2016, countries exporting catfish to the United States and intending to continue those exports after September 1, 2017, were required to provide FSIS with documentation demonstrating that their food safety systems generally comply with U.S. requirements, including the use

[36] The transition period also provides time for domestic catfish processors to bring their operations into full compliance with FSIS regulations.

of HACCP.[37] In addition, by that date, these countries were to provide FSIS with a list of processors intending to continue exporting catfish to the United States. As of June 2017, 13 countries submitted the documentation required to continue exporting catfish to the United States during the transition period.

In April 2016, FSIS assumed responsibility for inspecting imported catfish at ports of entry. FSIS refers to this as re-inspection, and it generally included (1) physical examinations and (2) collecting samples for testing for unsafe drug residues or other contaminants. During the transition, FSIS targeted shipments for re-inspection on at least a quarterly basis.[38] As part of the re-inspection, FSIS tested for unsafe drug residues, including for all drugs of concern to the agency and their corresponding MRLs. FSIS used, in part, an MRM that tested for 61 drugs. In cases in which an MRL was exceeded, FSIS considered the imported catfish to be adulterated and subject to refusal. For drugs approved by FDA for use in catfish, FSIS's MRLs are based on FDA's tolerances for these drugs. For unapproved drugs, FSIS's MRLs are based on levels determined by the agency.[39]

According to FSIS re-inspection data, from May 1, 2016, through July 9, 2017, the agency collected and tested 382 samples from 195 shipments of imported catfish for unsafe drug residues. Those shipments totaling over 4.4 million pounds came from 57 processors. FSIS found unsafe drug residues in 20 of the shipments and refused their entry into the United States. Together, these 20 shipments included about 422,000 pounds of catfish and came primarily from processors either in China or Vietnam.

---

[37] During the transition, FSIS required foreign countries to show that they had laws or other legal measures in place that ensure that their respective catfish processors comply with FDA HACCP regulations.

[38] According to FSIS documents, the criteria for selecting shipments for reinspection were based, in part, on the exporting country and volume of product exported by catfish processors in that country.

[39] FSIS refers to these levels as minimum levels of applicability and any catfish with drug residues above these levels is considered adulterated.

The drugs involved included dyes used as anti-fungal agents and antibiotics that are not approved for use in catfish in the United States.[40]

Through Its Equivalence Determination Process, FSIS Relies on Other Countries to Ensure the Safety of the Food They Export to the United States

As discussed, under FSIS's December 2015 final rule for its catfish inspection program,[41] a country wanting to continue exporting catfish to the United States after September 1, 2017, is to submit specified documentation and information to FSIS by that date. According to an FSIS document, the agency will use a country's documentation and information to begin assessing whether the country's sanitary measures for catfish provide an equivalent level of public health protection as found in the United States. According to FSIS documents, a country is expected to provide information about its food safety laws and regulations; inspection procedures, including manuals and directives; enforcement and compliance programs and policies; and inspection training programs.[42] According to FSIS documents, the equivalence determination process may take several years to complete.[43] During that time, FSIS may ask the country for additional information, as needed, and FSIS officials are to conduct an on-site (in-country) audit, and in many cases more than one to verify the information provided by the country. FSIS conducts an on-site audit at least once during an initial equivalence determination and again periodically during subsequent verification audits. According to agency

---

[40] The drugs found by FSIS included gentian violet, malachite green, nitrofurans, and enrofloxacin. Gentian violet and malachite green are dyes effective against fungal infections. Nitrofurans and enrofloxacin are antibiotics.

[41] 80 Fed. Reg. 75,590 (Dec. 2, 2015).

[42] FSIS expects that the country's competent government authority for food safety will address the Self-Reporting Tool questions according to these documents.

[43] The explanatory statement associated with the 2017 Consolidated Appropriations Act directs FSIS to complete an equivalence determination for the catfish inspection program in a foreign country that exports catfish and catfish products to the United States not later than 180 days after the end of the 18-month transition period specified in FSIS's final catfish inspection rule—March 1, 2018.

officials, FSIS will conduct periodic on-site audits in countries with equivalence determinations to verify that their food safety systems remain equivalent. FSIS uses a data analysis tool, known as the Country Performance Algorithm, to prioritize the countries subject to these audits. The agency also publishes the related audit reports on its website. In the course of an on-site audit, FSIS officials generally visit government offices, commercial food processing facilities, and food testing laboratories in the country.

In essence, FSIS intends to use the same equivalence determination process for imported catfish that it currently uses for imported meat, poultry, and processed egg products, according to agency officials. As part of the equivalence determination, FSIS also requires a foreign country to develop a residue monitoring plan that includes testing for drug residues. According to FSIS documents, these plans must include, in part, random sampling of food products at slaughter and the use of approved sampling and testing analytical methods. We also discussed, in part, FSIS's equivalence determination process in an April 2017 report.[44]

In October 2016, FSIS finalized its Self-Reporting Tool (SRT), a standardized questionnaire that the agency provides to countries that request an equivalence determination for catfish. The SRT describes the types of documentation and information that the country must provide to FSIS for the agency to initiate the equivalence determination process. In general, this documentation and information relates to the country's food safety system and regulatory infrastructure. For example, the SRT asks for documentation and information on the following six equivalence components of a country's food safety system

---

[44] GAO, *Foot-and-Mouth Disease: USDA's Evaluations of Foreign Animal Health Systems Could Benefit from Better Guidance and Greater Transparency*, GAO-17-373 (Washington, D.C.: Apr. 28, 2017). GAO recommended that USDA (1) clarify its guidance on how staff should document analysis of a foreign country's animal health system and the results of in-country visits to verify information; (2) complete its efforts to develop an information management system to better store, organize, and manage documentation gathered about a foreign country's animal health system; and (3) develop guidance promoting greater transparency in risk analysis reports in accordance with the quality information guidelines issued by USDA's Chief Information Officer and guidance from the Office of Management and Budget. USDA concurred with these recommendations.

- government oversight,
- statutory authority for food safety and other consumer protection regulations,
- sanitation,
- HACCP compliance,[45]
- chemical residue testing programs, and
- microbiological testing programs.

According to FSIS officials, as of May 2017, 10 countries had requested an equivalence determination and FSIS provided them with the SRT. These countries were Bangladesh, China, the Dominican Republic, El Salvador, Guyana, Jamaica, India, Nigeria, Thailand, and Vietnam. Furthermore, as of June 2017, Guyana and Thailand had provided sufficient information for FSIS to begin the equivalence determination process.

According to FSIS documents, once it has completed its review and analysis of a country's documentation and information, including the results of its own on-site audits, if deemed satisfactory, FSIS is to then publish a proposed rule in the *Federal Register* indicating its intention to add the country to FSIS's list of countries eligible to export a particular food commodity to the United States. After receiving and reviewing any public comments on the proposed rule, FSIS will make a final equivalence determination decision and, as appropriate, publish a final rule in the *Federal Register* regarding the country's eligibility to export the commodity to the United States. As discussed, the equivalence determination process can take several years from the time a country completes and submits the SRT until FSIS makes its final determination. According to agency officials and FSIS documentation, FSIS will allow a country that submitted an SRT to continue exporting catfish to the United States pending completion of its equivalence determination as long as that

---

[45] According to FSIS documents, following the transition, foreign governments will have to show that they require their catfish processors to use a HACCP system comparable to FSIS's HACCP requirements as part of the equivalence determination process.

country continues to respond to FSIS's requests for additional information within the timeframe indicated.

According to FSIS officials and documents, after an equivalence determination is completed, the country reviewed is authorized to export the related food product to the United States. Thereafter, on an annual basis, FSIS expects the country's government to review and update its SRT responses, as appropriate, and notify FSIS of any changes to the country's food safety system. SRT revisions may be necessary because of country-initiated changes, or revisions may be needed because of new policies adopted by the United States. FSIS will also require the country to submit, on an annual basis, specific documents, including an updated list of all certified food processing facilities eligible to export to the United States, and an updated description of the country's residue monitoring plan, including the previous year's test results and any actions taken in response to unsafe drug residue test results.

According to FSIS officials, after the transition period, all shipments of imported catfish from countries that the agency has determined to have an equivalent food safety system will have to be presented to FSIS inspectors for reinspection. In addition, a subset of reinspected shipments will be sampled. According to FSIS officials, after the 18-month transition period, the agency will update the sampling program based on sampling results and findings from the transition period. Furthermore, according to agency officials and FSIS documents, shipments of imported catfish may be subject to the same three levels of sampling used for imported meat and poultry: (1) normal sampling, which is based on random sampling; (2) increased sampling, which is above-normal sampling resulting from an agency management decision; and (3) intensified sampling, which is additional sampling undertaken when a previous sample failed to meet U.S. requirements, such as drug residues that are above the corresponding MRL. According to FSIS officials and documents, after equivalence determinations have been done, the intent of the reinspection process is to verify the effectiveness of the foreign country's food safety system, not to evaluate the performance of individual catfish processing facilities or as the primary effort/point to identify unsafe drug residues in imported catfish.

FSIS Could Strengthen Its Efforts to Ensure the Safety of Imported of Catfish

As discussed, during FSIS's on-site audits, which the agency conducts at least once to verify the accuracy of the documentation and information a foreign country seeking an equivalence determination has provided in an SRT, FSIS generally visits government offices, commercial food processing facilities, and food testing laboratories in the foreign country. However, FSIS has not made farm visits—where catfish exported to the United States are grown and where drugs are potentially first introduced into catfish—a routine part of initial equivalence determination and verification on-site audits. In addition, although FSIS will require foreign countries to develop and implement residue monitoring plans, the agency will not require these countries to test for drugs of concern to FSIS. In contrast, the EU requires countries to test for drugs of concern to the EU.

The 2014 Farm Bill directs USDA to consider the conditions under which catfish are raised and transported to the processing establishment for examination and inspection. According to agency officials, this directive allows FSIS to visit catfish farms in exporting countries as part of its on-site audits related to an initial equivalence determination or a subsequent verification audit. However, these officials said FSIS has not made farm visits a routine part of initial equivalence determination and verification on-site audits, because the agency has not yet decided whether to make visits to fish farms routine like visits to foreign laboratories or processing facilities.

It is at the fish farms where drugs are first introduced and used. Other seafood regulators conduct such visits to fish farms, including FDA and the EU. For example, according to FDA's procedures for conducting country assessments, visiting fish farms is a critical element in evaluating a country's seafood safety oversight program because farms are where the potential hazard of unsafe drug residues originates. At such farms, FDA officials said they review (1) preventive controls to guard against unapproved drug use; (2) water quality or other factors that may lead farmers to use drugs; (3) drug use records, including drug withdrawal times[46] (4) information on countries' fish farm oversight programs (e.g., registration, inspection, and sampling); (5) data that can be compared to seafood processor records to determine the accuracy and reliability of these records; and (6) biosecurity measures meant to prevent the spread of disease and keep fish healthy. According to FDA documents, FDA officials visit farms that supply the products to the seafood processing facilities scheduled to be visited by FDA during the assessment. This is done to verify the controls stated in the processors' seafood HACCP plans. Further, by visiting fish farms, FDA gathers useful information about any foreign government oversight of the farms, as well as the drugs that are actually being used on the farms in that country, according to agency documents. For example, as part of FDA's country assessment for Ecuador, FDA staff visiting two fish farms in that country discovered that the farms were using an antibiotic on shrimp that FDA had not approved for that product.[47]

The EU also visits fish farms in foreign countries during its initial equivalence determinations and periodic verification audits. According to EU officials, by visiting these farms and other locations in a country, the EU audit team occasionally learns of drugs being used by farms that are not included in the country's residue monitoring plan. For example, during the EU's 2012 verification audit of Ecuador, EU inspectors visited fish farms and found them using a hormone that the Ecuadorian government could not verify it had approved for use in aquaculture. In addition, the information on the drug label did not indicate that it was authorized for use in Ecuador. Furthermore, one of the farms visited had barrels of an

---

[46] The withdrawal time is the period following the last treatment with a drug during which an animal may not be offered for slaughter and products from the animal, such as milk and eggs, may not be offered for sale.

[47] However, Ecuador was not under any obligation or requirement to take any action on this matter because it does not have a formal agreement with the United States.

antibiotic that did not bear numbers or labels, indicating that the antibiotic was authorized for use in Ecuador.

It is not clear how FSIS could consider the conditions under which imported catfish are raised, as directed by the 2014 Farm Bill, without visiting foreign catfish farms. In addition, the agency will already have its inspectors in the foreign country for an on-site audit. Without visiting at least a sample of farms whose catfish are exported to the United States, FSIS may be missing an opportunity to fully understand the conditions under which the catfish are being raised. For example, FSIS could choose to visit farms that supply catfish to seafood processing facilities that the agency plans to visit during its on-site audits, like FDA does.

FSIS also does not plan to require countries exporting catfish to the United States to test for drugs of concern to FSIS as part of a country's drug residue monitoring plan, according to agency officials. Instead, as part of an equivalence determination, FSIS will require a country to have a national residue monitoring plan. Further, the officials said that FSIS will expect this plan to include such information as the (1) chemicals, including drugs, that will be tested for; (2) proposed number of samples to be taken; (3) testing methods to be used for screening and confirming the presence of chemicals, including drugs; (4) MRLs to be used; (5) chemicals, including drugs, newly approved or banned in the past year; and (6) corrective actions to be taken when a residue violation is found. However, while agency officials said that FSIS expects another country to provide an overall level of sanitary protection similar to that in the United States, the country has flexibility regarding its regulatory practices, including the specific drugs included in its national residue monitoring plan.

Even if a country allows its catfish farmers to use drugs not approved for use in the United States, that country's catfish exports must still comply with FSIS's requirements (i.e., MRLs) for allowable drug residues.[48] As noted by agency officials, FSIS expects imported catfish to be subject to the same level of scrutiny as domestic catfish. However, while FSIS will

---

[48] FDA has approved nine drugs for use in domestic catfish. Catfish farmers in other countries are able to use all drugs approved by their governments, regardless of whether those drugs are approved by FDA.

test domestic catfish for at least 61 drugs using its MRM and other methods, it is not clear how many drugs or which drugs other countries will test for in catfish exported to the United States because FSIS will not have such specific requirements for a foreign country's residue monitoring plan after the transition period. Further, FSIS will assess the safety of drug residues found in domestic catfish against MRLs determined by the agency, although other countries may use their own MRLs in their drug residue monitoring plans. For example, Vietnam, the largest exporter of catfish to the United States, developed its drug residue monitoring plan and MRLs to meet the EU's requirements. In comparing FSIS's MRLs to Vietnam's levels for 14 drugs that could be used in catfish, we found 9 drugs for which Vietnam had higher MRLs than FSIS's. For one antibiotic, the residue level that Vietnam considers a violation is 30 times greater than FSIS's level.

The first time imported catfish may be subject to drug residue testing similar to the testing done of domestic catfish, including the same 61 drugs and associated MRLs, is likely to be when, and if, the imported catfish is selected for sampling by FSIS in U.S. ports of entry. As a result, even after equivalence determinations are completed, FSIS's sampling and testing of imported catfish shipments will become the primary mechanism for identifying the potential presence of unsafe residues of drugs of concern to FSIS. According to FSIS officials, as of July 2017, the agency has not yet documented the size of the samples that it will test for during fiscal year 2018.

In contrast, the EU requires countries exporting seafood to the EU to include drugs of concern to the EU in their residue monitoring plans. Although FSIS officials said that the agency expects imported catfish to be subject to the same level of scrutiny as domestic catfish and that reinspection is to verify the effectiveness of the exporting country's food safety system, the agency is not requiring that the foreign countries test for the drugs of concern to FSIS and the corresponding MRLs, which would qualify as the initial foreign country inspection. By having countries exporting catfish to the United States include in their national residue monitoring plans the drugs of concern to FSIS and the corresponding

MRLs, as a precondition for an equivalence determination, FSIS would have better assurance that U.S. food safety requirements were being met and that U.S. consumers were being protected.

FDA and FSIS Took Key Steps to Coordinate the Transfer of Catfish Oversight, but They Have Not Fully Coordinated on Drug Residue Testing Methods

FDA and FSIS took steps to accomplish the transfer of catfish oversight from FDA to FSIS, as called for in the 2014 memorandum of understanding (MOU) that both agencies signed. However, they generally have not coordinated on drug residue testing methods, resulting, in some cases, in differences in MRLs—specifically for unapproved drugs. FDA and FSIS agreed in their 1984 MOU to exchange information on their drug testing methods and ensure the comparability of their results.

FDA and FSIS Took a Number of Steps to Coordinate the Transfer of Catfish Oversight

In April 2014, FDA and FSIS entered into an MOU to improve interagency cooperation on food safety and fraud prevention and maximize the effectiveness of personnel and resources related to examination and inspection, and to plan for an orderly transition of catfish safety oversight from FDA to FSIS. The 2014 Farm Bill, which required the agencies to develop this MOU, directed that the MOU should build upon any prior agreements between the agencies to improve interagency cooperation. Pursuant to this MOU, FDA and FSIS took a number of steps to coordinate the transfer, including the following:

- According to FDA officials, in January 2015, FDA established an internal FDA work group that compiled a list of questions and answers for FDA and FSIS management officials on topics related to the transfer.

- According to FSIS officials, FDA and FSIS established a transition work group comprising of FDA and FSIS management officials and subject matter experts. According to FSIS officials, members of this workgroup met periodically to discuss issues pertaining to the transition of jurisdiction and regulatory oversight of catfish from FDA to FSIS. According to FDA officials, the first transition group conference call was held on May 19, 2015.
- According to FSIS officials, from May 2015 through June 2016, FDA and FSIS held three teleconferences to coordinate activities related to the transfer. Specifically, agency officials discussed inspection methodologies, sampling, and laboratory testing; FDA's previous strategy for regulatory oversight of imported catfish, including relevant FDA import alerts; and sharing of FDA's knowledge of fish farms and processing facilities in the United States that handle catfish.
- According to FSIS officials, before publishing the final rule for its catfish oversight program in December 2015, FSIS provided a draft of this rule to FDA for its review and comment.
- FDA and FSIS agreed—as FSIS assumed responsibility for domestic catfish oversight on March 1, 2016, and imported catfish oversight on April 15, 2016—that there would be no duplication of the inspection and testing of catfish between the two agencies. Further, FDA agreed not to inspect domestic or foreign catfish processing facilities or test catfish products unless FSIS requested such inspections.

Further, after the agencies took these steps and the transfer was completed, in discussions with FDA and FSIS officials in January 2017, we mentioned that FDA appeared to have information that might be useful to FSIS in carrying out its catfish oversight program that FDA had not shared with FSIS. We suggested that FDA share with FSIS its past (1) drug residue testing data for catfish, (2) catfish processor inspection reports, and

(3) foreign country assessments for catfish-exporting countries. For example, we suggested that information in FDA's foreign country assessments might be useful to FSIS in doing equivalence determinations for some of the same countries. After these meetings, FSIS requested, and FDA provided, this additional information.

FDA and FSIS have not Fully Coordinated on Drug Residue Testing Methods

Although FDA and FSIS took steps to implement the 2014 MOU and ensure the smooth transfer of catfish oversight responsibility from FDA to FSIS, the agencies have not fully coordinated on drug residue testing methods. For example, the agencies generally did not coordinate in developing drug residue testing methods and the corresponding MRLs—specifically, for unapproved drugs—that define unsafe drug residues in imported seafood, including catfish. As a result, the agencies are not leveraging each other's knowledge and resources to develop drug residue testing methods.

As discussed, FDA and FSIS staff held three teleconferences from May 2015 through June 2016 to discuss the transfer of catfish oversight to FSIS. According to agency officials, the sampling and testing of catfish for drug residues was among the topics discussed, including the drug testing methods being used by each agency. Nevertheless, during the course of our work, we found examples where FDA and FSIS officials, including program directors and laboratory officials, said that they were unaware of the other agency's development and use of more efficient methods to detect drug residues in seafood, including catfish. These examples include the following:

- FDA officials first learned from us in May 2016 about an MRM used by FSIS that could identify a significantly greater number of drug residues present in food, including catfish, than the MRM being used by FDA for other seafood. Specifically, at that time,

FSIS's MRM could identify 61 drug residues, and FDA's MRM could identify 26 residues.

- FDA had not informed FSIS that it was developing a new MRM for testing shrimp, although according to agency officials FDA also expected to adapt the MRM for use in other seafood, including finfish such as catfish. FSIS officials said that they became aware of FDA's work on this new MRM after hearing FDA officials discuss it at a public workshop (the North American Chemical Residue Workshop) in July 2016.
- At the time of our work, FDA was developing two new MRMs that would test for, among other drugs, the antibiotics flumequine, naladixic acid, and oxolinic acid—drugs that do not have FDA approval for use in aquaculture. Pending development of these MRMs, FDA had been using other methods—such as the MRM that tests for 26 drugs—to test for these drugs, including in catfish when FDA had this responsibility. According to FDA officials, the agency tested for these drugs because they were of high enforcement priority based on human food safety concerns and on the extent of use in the aquaculture industry. However, FSIS officials told us in May 2016 that FDA had not informed them of its efforts to develop these two new MRMs, and FSIS did not have its own method to test for these drugs, meaning imported catfish had not been tested for these antibiotics since FSIS assumed responsibility for their safety in April 2016.

As these examples illustrate, neither FDA nor FSIS was aware of MRMs or other testing methods that the other agency was using or developing that might have been helpful for carrying out its oversight program. As we have previously reported, when responsibilities cut across more than one federal agency—as they do for the safety of imported seafood—it is important for agencies to work collaboratively.[49] Taking into

[49] GAO, *2015 Annual Report: Additional Opportunities to Reduce Fragmentation, Overlap, and Duplication and Achieve Other Financial Benefits*, GAO-15-404SP (Washington, D.C.: Apr. 14, 2015), GAO-06-15.

account the nation's long-range fiscal challenges, we noted that the federal government must identify ways to deliver results more efficiently and in a way that is consistent with its multiple demands and limited resources. We also identified key practices that can help enhance and sustain federal agencies' collaboration, including identifying and addressing needs by leveraging resources across agencies to support the common outcome. Moreover, in 1984, FDA, FSIS, and the Environmental Protection Agency entered into an MOU committing the agencies, in part, to coordinate their regulatory activities concerning the presence of drug and other residues in food. More specifically, under this MOU, the agencies agreed to exchange information related to analytical methods for identifying and quantifying residues of drugs, pesticides, and environmental contaminants in food, and to cooperate in developing and implementing analytical and statistical methodologies to ensure comparability of results in the examination of food. Thus, by not coordinating on testing methods, FDA and FSIS have not fully met the terms outlined in the 1984 MOU.

Lack of coordination in developing drug residue testing methods has resulted, in some cases, in differences in MRLs for unapproved drugs. These are the levels or thresholds at which the agencies will take regulatory action. In some cases, particularly for an unapproved drug, the MRL may correspond to the limits of detection associated with an agency's particular testing method. In the course of our work, we noted a number of cases where the agencies were using different MRLs for the same unapproved drug. For example, for one antibiotic, the MRL that FSIS considers unsafe is 20 times higher than FDA's level, yet FDA considers anything above its MRL to be a human health concern. In essence, this means that FSIS does not consider catfish with residue levels greater than FDA's MRL, but less than FSIS's MRL, to be adulterated. However, when FDA had oversight of catfish, the same level would have rendered the product adulterated.[50] We discussed this and other cases with FSIS officials, and they said that FSIS uses FDA's tolerances for drugs that FDA has approved. However, FSIS developed its own MRLs for unapproved

---

[50] This level would render other types of finfish, which FDA still regulates, adulterated as well. Catfish is classified as a finfish.

drugs, as according to FSIS officials, the agency has the discretion to do so using its own testing methods.

According to FSIS officials, the agency was not aware of FDA's MRLs for unapproved drugs. FSIS officials acknowledged that the agency had not requested information from FDA on its MRLs for unapproved drugs used on finfish, including catfish. For their part, FDA officials said that they do not make public their MRLs for unapproved drugs out of concern that some fish farmers may use these drugs if they believe they can do so in a way that results in residues below FDA's MRLs. Under federal standards for internal control, management should communicate information externally through reporting lines so that external parties can help achieve its objectives. Without coordinating and communicating on their development of drug residue testing methods and corresponding MRLs for imported seafood, including catfish, the agencies do not have reasonable assurance that they are taking a consistent approach to protecting consumer safety from unsafe drug residues.

CONCLUSION

FDA and FSIS face difficult challenges in ensuring the safety of the U.S. food supply, particularly as that food supply increasingly includes imported foods such as seafood. Because much of imported seafood is raised in confined conditions on farms, drugs are used to prevent or treat disease and increase survival rates. According to FDA, residues of some drugs can cause cancer or allergic reactions when consumed by humans. In addition, some drugs administered to food-producing animals may cause bacteria of human health concern to become resistant to antibiotics used in humans. It is therefore important that federal oversight is effective in ensuring that seafood is free of unsafe drug residues

As required in FSMA, FDA developed a plan in 2013 to expand the capacity of foreign governments and their respective food industries in countries that export foods to the United States. The plan was to include recommendations for formal agreements with other countries that would include provisions to place greater responsibility on these countries for the safety of their seafood exports. However, while FDA has used such agreements for addressing pathogen hazards in molluscan shellfish, it has not done so with respect to drug residues in seafood and has no plans to do so. Without pursuing formal agreements with countries exporting seafood to the United States that commit these countries to test for drugs of concern to FDA and the corresponding MRLs that the agency established for these drugs, FDA will not have reasonable assurance that imported seafood does not contain unsafe drug residues.

FSIS has not decided whether to include, as part of an initial equivalence determination or subsequent verification audits, visits to any foreign catfish farms as part of its on-site audit in another country. However, the 2014 Farm Bill directs FSIS to consider, in part, the conditions under which catfish are raised, and catfish farms are the place where drugs are introduced. Without visiting at least a sample of farms whose catfish are exported to the United States, such as the farms that supply catfish to the seafood processing facilities that FSIS plans to visit during its on-site audits, FSIS may be missing an opportunity to fully understand the conditions under which the catfish are being raised. In addition, FSIS does not plan to require countries exporting catfish to the United States to test for drugs of concern to FSIS as part of their drug residue monitoring plans. By having countries exporting catfish to the United States include in their national residue monitoring plans the drugs of concern to FSIS and the corresponding MRLs, as a precondition for equivalence determinations, FSIS would have better assurance that U.S. food safety requirements were being met and that U.S. consumers were being protected.

Finally, FDA and FSIS are independently developing drug testing methods and MRLs for use in seafood, and lack of coordination and communication in developing drug residue testing methods has resulted, in some cases, in differences in MRLs—specifically, for unapproved drugs. Without coordinating and communicating on their development of drug residue testing methods and corresponding MRLs for imported seafood, including catfish, the agencies do not have reasonable assurance that they are taking a consistent approach to ensuring consumer safety from unsafe drug residues.

RECOMMENDATIONS FOR EXECUTIVE ACTION

We are making a total of five recommendations, including two to FDA and three to FSIS:

The Commissioner of FDA should pursue formal agreements with countries exporting seafood to the United States that commit these countries to test for drugs of concern to FDA and the corresponding MRLs that FDA established for these drugs. (Recommendation 1)

The Administrator of FSIS should ensure that agency staff doing an on- site audit in another country for an equivalence determination visit at least a sample of farms whose catfish are exported to the United States to determine the conditions under which the catfish are being raised, including the drugs being used. (Recommendation 2)

The Administrator of FSIS should require as part of an equivalence determination that countries exporting catfish to the United States include in their residue monitoring plans the drugs of concern to FSIS and the corresponding maximum residue levels. (Recommendation 3)

The Commissioner of FDA should coordinate and communicate with FSIS in developing drug residue testing methods and corresponding maximum residue levels for imported seafood that may also be applicable to imported catfish. (Recommendation 4)

The Administrator of FSIS should coordinate and communicate with FDA in developing drug residue testing methods and corresponding maximum residue levels for imported catfish that may also be applicable to other imported seafood. (Recommendation 5)

AGENCY COMMENTS AND OUR EVALUATION

We provided a draft of this chapter to USDA and the Departments of Commerce, Health and Human Services, and Homeland Security for their review and comment. In written comments, Health and Human Services's FDA agreed with one of the recommendations and partially agreed with the other. FDA also provided technical comments, which we incorporated as appropriate. In written comments, USDA's FSIS partially agreed with two of the recommendations and stated that its current policy already addresses the third recommendation. Copies of Health and Human Services's and USDA's comments are presented in appendixes III and IV, respectively. In an email, the Department of Homeland Security's GAO-Office of Inspector General Liaison Office stated that, because there were no recommendations directed to the department, it would forego a formal management response letter but provided one technical comment, which we incorporated. Likewise in an e-mail, the Department of Commerce's National Oceanic and Atmospheric Administration Audits Office stated that the agency did not have any technical comments.

In its written comments, FDA agreed with our recommendation that it should coordinate and communicate with FSIS in developing drug residue testing methods and corresponding maximum residue levels for imported seafood that may also be applicable to imported catfish. FDA stated that it has a process in place to notify FSIS of new tolerances and changes in tolerances for FSIS-regulated products and that it will extend this process to notify FSIS of concentrations of specific unapproved drugs in catfish over which FDA has public health concerns. In addition, FDA stated that it has contacted FSIS about rejoining the quarterly meetings FDA holds on aquaculture method prioritization and development.

FDA partially agreed with our recommendation that it should pursue formal agreements with countries exporting seafood to the United States that commit these countries to test for drugs of concern to FDA and the corresponding MRLs that FDA established for these drugs. FDA stated that, while it had not received any requests to establish this type of arrangement, it concurs that the agency could explore pursuing such arrangements. FDA also stated that factors outside its control that could limit robust implementation of this recommendation include the country's ability and readiness to comply with the requirements necessary to have a successful arrangement. FDA added that applicable test methods, analytical capacity and adequate government oversight would be among essential criteria. We recognize that there are external factors that could affect FDA's implementation of this recommendation, but we do not believe such factors should prevent FDA from pursuing formal agreements with other countries related to testing seafood for drugs of concern and the related MRLs for these drugs, as these factors have not hindered the EU and other countries that have successfully pursued such agreements.

In its written comments, FSIS stated that the draft report contains a few either misleading or inaccurate statements that it believes we use to support our recommendations directed at USDA. For example, FSIS points to our statement that "FSIS has not made farm visits a routine part of initial equivalence determinations and verification on-site audits...." According to FSIS, this statement is somewhat misleading, because FSIS has not made any initial equivalence determinations for foreign catfish fish inspection systems;[51] therefore, FSIS has not yet had the opportunity to conduct on-site verification audits of foreign catfish inspection programs for initial equivalence determinations. FSIS stated that it partially agrees with the recommendation that FSIS visit at least a sample of farms whose catfish are exported to the United States to determine the conditions under which the fish are raised. We are aware that FSIS has not yet conducted on-site audits, and we discuss in our report the timing of those audits once the agency receives the required information and documentation from foreign governments. However, we nonetheless believe it is important that FSIS visit at least a sample of catfish farms as a routine matter during its equivalence determination on-site audits, instead of relying on a review of documentation describing a foreign country's fish farm oversight program.

[51] FSIS stated that it is more accurate to use the term Siluriformes fish throughout the report, as it is the order of fish under FSIS jurisdiction, and that by using the term "catfish" throughout the report, it appears to invalidate the clarification in the 2014 amendment (Sec. 12106) in the Agricultural Act of 2014, which GAO may not have intended. Consistent with all of our reports on this issue, we use the term catfish to refer to Siluriformes fish.

We believe that FSIS can best understand the conditions under which catfish are raised and obtain information about the drugs actually used on farms through in-person visits to these farms, as FDA and the EU do as part of their seafood oversight efforts. As noted in our report, FSIS will already have its inspectors in the foreign country for on-site audits, so these individuals could also visit the farms that supply catfish to the seafood processing facilities that FSIS plans to visit during its on-site audits. FSIS could also independently verify any foreign country information about its catfish farm oversight program through visits to catfish farms.

FSIS stated that its current policy already addresses our recommendation that it require, as part of an equivalence determination, that countries exporting catfish to the United States include in their residue monitoring plans the drugs of concern to FSIS and the corresponding maximum residue levels. FSIS identifies as misleading or inaccurate our statement that, "while FSIS will test domestic catfish for at least 61 drugs using its multi-residue method and other methods, it is not clear how many drugs or which drugs other countries will test for in catfish exported to the United States." According to FSIS, this statement is incorrect, as foreign countries are required to provide drug testing information to the agency as part of the Self-Reporting Tool (SRT), which is a requirement for an initial or ongoing equivalence determination. In addition, FSIS stated that it will also have access to information on a foreign country's testing for drug residues through the SRT and the country's residue monitoring plan. Further, after reviewing those plans, according to FSIS, the agency can request changes to the plan for testing for drug residues. We disagree that FSIS policy already addresses our recommendation. In our report, we explain the type of information that foreign countries will provide in the SRT as well as the information that foreign countries will be required to include in their residue monitoring plan. However, a foreign country will decide what drugs it will test for. FSIS could question the design of the residue monitoring plan if it identified unsafe drug residues based on its own testing. Nevertheless, FSIS does not plan to test all catfish imports because its reinspection program is not designed to be the primary means by which the agency identifies unsafe drug residues in imported catfish. Rather than address potential testing gaps in a foreign country's residue monitoring program piecemeal, FSIS should require that foreign countries test for all drugs of concern to FSIS at the outset of the equivalence determination, thus ensuring that foreign countries are demonstrating that their measures are as effective as FSIS's in addressing the safety of imported catfish.

FSIS partially agreed with our recommendation that FSIS should coordinate and communicate with FDA in developing drug residue testing methods and corresponding MRLs for imported catfish that may also be applicable to other imported seafood. In its comments, FSIS noted that while we found examples where FDA and FSIS officials, including program directors and laboratory officials, were unaware of the other agency's development and use of more efficient methods to detect drug residues in seafood, including catfish, they believe that the characterization of this finding is overstated. FSIS stated that while our audit work may have found examples where officials were unaware of a specific activity, the U.S. Agricultural Research Service, FSIS, and FDA regularly share information related to analytical methods for identifying and quantifying residues of drugs, pesticides, and environmental contaminants. Nevertheless, FSIS stated that it fully intends to implement the provisions of the MOU with FDA on coordinating on testing methods and that it also intends to enhance residue testing coordination through other interagency mechanisms as well, such as the Surveillance Advisory Team and the Interagency Residue Control Group.

As agreed with your office, unless you publicly announce the contents earlier, we plan no further distribution of this chapter until 30 days from the report date. At that time, we will send copies to the appropriate congressional committees; the Secretaries of Agriculture, Commerce, Health and Human Services, and Homeland Security; and other interested parties. Sincerely yours,

Steve Morris
Director, Natural Resources and Environment

APPENDIX I: OBJECTIVES, SCOPE AND METHODOLOGY

This chapter addresses how the Food and Drug Administration (FDA) and U.S. Department of Agriculture's (USDA) Food Safety and Inspection Service (FSIS) ensure the safety of imported seafood from unsafe drug residues. Specifically, this chapter examines (1) how FDA helps ensure the safety of imported seafood from unsafe drug residues and ways the agency could strengthen its efforts, (2) how FSIS helps ensure the safety of imported catfish from unsafe drug residues and ways the agency could strengthen its efforts, and (3) the extent to which FDA and FSIS coordinate their oversight efforts. Unsafe drug residues may include residues

- from drugs unapproved for aquaculture use in the United States,
- of approved drugs that exceed allowed levels, or
- of drugs that are approved for use on one seafood species but are being inappropriately used on another seafood species.

To review how FDA helps ensure the safety of imported seafood from unsafe drug residues, we examined FDA regulations, including Hazard Analysis and Critical Control Point (HACCP) regulations. We also reviewed information on FDA's primary oversight mechanism—HACCP inspections of seafood importers and foreign country processing facilities—and its seafood import sampling program, including information on the major components and requirements of these mechanisms. We reviewed relevant sections of FDA guidance manuals, including its *Compliance Program Guidance Manual, Regulatory Procedures Manual, Office of Regulatory Affairs Laboratory Manual, and Fish and Fishery Products Hazards and Controls Guidance - Fourth Edition.* We also reviewed plans and other documents detailing FDA's drug residue sampling program and procedures for its import alert and refusal processes. We analyzed FDA mechanisms for protecting imported seafood from violative drug residues. Specifically, we analyzed 74 randomly selected foreign facility (e.g., processing) inspection reports from a total of 318 reports for fiscal years 2013 through 2015. We also analyzed 9 non generalizable, randomly selected importer inspection reports out of 232 total reports for fiscal year 2015.[52]

To examine the ways FDA could strengthen its efforts to ensure the safety of imported seafood from unsafe drug residues, we reviewed FDA's use of agreements with foreign countries that address FDA requirements regarding drug residues in farmed seafood and their importance, given limitations in other activities FDA undertakes to ensure the safety of imported seafood. We reviewed legal and planning documents regarding the use of agreements by FDA, including the FDA Food Safety Modernization Act and FDA's 2013 International Food Safety Capacity-Building Plan. We reviewed a 2012 agreement with Mexico intended to ensure that all molluscan shellfish exported to the United States from Mexico are safe. In addition, we reviewed FDA's foreign country assessment criteria and reports for Ecuador (2013), India (2010), Malaysia (2011), the Philippines (2015), and Vietnam (2012) and how any FDA-identified deficiencies were resolved. FDA conducts foreign country assessments to provide the agency with a broad view of a foreign country's industry and regulatory infrastructure capacity to control aquaculture drugs. We also compared FDA activities to the activities FSIS and the European Union (EU) undertake as part of their equivalence determinations to determine if there are any elements of their activities that if included in FDA activities, have the potential to enhance FDA's oversight process for imported seafood.

In addition, we analyzed FDA's import refusal data and data on FDA's import drug residue test results for fiscal years 2010 to 2015. For FDA's import drug residue test data, we corrected the product codes for some entries for which the code did not match the product description and two analysts reviewed any changes made to FDA records to ensure that the revised codes were correct. We also reviewed controls for the systems that house these data and interviewed FDA officials regarding these controls

---

[52] One of the 10 reports for New Zealand was omitted because it is not required to identify affirmative steps because it has a systems recognition agreement with the United States. Systems recognition means that FDA identified the country as having a comparable food safety system to the United States. As such, it is subject to different requirements than what the agency has for seafood imports from other countries, according to FDA officials.

and determined that the data were sufficiently reliable for the purposes of our reporting objectives. We analyzed the National Oceanic and Atmospheric Administration's (NOAA) seafood import volume data for 2010 through 2015 that it acquired from the Department of Commerce and U.S. Customs and Border Protection (CBP) to enable us to identify how many millions of pounds were imported for each type of seafood so that we could compare this to FDA sampling data. We reviewed controls related to the reliability of these data and determined that they were sufficiently reliable for the purposes of our reporting objectives. Specifically, we reviewed a document on the data quality assessment of this information by the U.S. Census Bureau and compared the data to import data generated by another federal agency—USDA's Economic Research Service.

The final year selected for FDA data (foreign processing facility and importer inspection reports and import refusal data and data on FDA's import drug residue test results) and NOAA import volume data was the most recent year for which complete data were available at the time of our analysis.

To examine how FSIS helps ensure the safety of imported catfish from unsafe drug residues, we reviewed information on how the agency was implementing its catfish inspection program during the program's transition period from March 1, 2016, to September 1, 2017, and how it planned to implement its equivalence determination process for catfish. Specifically, for the transition period, we reviewed the 2015 final rule that established the catfish inspection program; FSIS notice and other information on the catfish inspection program requirements; sampling and testing guidance during the transition period, including Hold and Test Protocols; FSIS data on sampling and testing and the results; FSIS data on catfish import sampling from May 1, 2016, through July 9, 2017; and FSIS's *Chemistry Laboratory Guidebook,* including the Screening and Confirmation of Animal Drug Residues Method (CLG-MRM1.06). We also reviewed the agency's controls for the systems that house these data and determined that the data were sufficiently reliable for the purposes of our reporting objectives. For example, we interviewed FSIS officials regarding supervisory approvals of test results input into their system and controls included in their efforts to comply with international standards.

For FSIS's plans for implementing its equivalence determination process after the end of the transition period, we reviewed FSIS regulations, guidance, and other documentation related to FSIS's equivalence determination process as currently used for meat, poultry, and processed egg products. Specifically, we reviewed the steps FSIS takes to determine equivalence, including the Self-Reporting Tool that the agency provides to the foreign governments that ask for equivalence determinations, which contains the general information required from the foreign countries. We reviewed FSIS guidance on the agency's periodic audits of countries to verify continued equivalence and two reports resulting from these audits that FSIS conducted of countries that have already been determined to be equivalent for meat and poultry. We also reviewed FSIS's reinspection program for monitoring the effectiveness of exporting countries' inspection systems and overall food safety programs through imported product examinations and residue testing. We reviewed an example of how FSIS places responsibility on foreign governments to take corrective actions when the agency finds that imported products are adulterated because of contaminants such as violative drug residues. To examine the ways, if any, FSIS could strengthen its efforts to ensure the safety of imported catfish from unsafe drug residues, we reviewed FSIS plans for its on-site equivalence determinations and verification audits and compared these activities to the activities FDA and the EU undertake as part of their foreign country assessments and equivalence determinations, respectively, to determine if there are any elements of their activities that if included in FSIS activities, have the potential to enhance FSIS's equivalence determinations and verification on-site audits. We also reviewed FSIS's proposed requirements for foreign countries' residue monitoring plans. Lastly, we reviewed the EU's requirements for foreign countries' residue monitoring plans as well to determine whether there were any specific EU requirements that could enhance FSIS's residue monitoring plan requirements. We reviewed the EU's equivalence determination process and, in particular, its seafood import program to determine whether its practices for ensuring the safety of seafood imports have the potential to enhance the U.S. agencies' practices. We discussed the EU equivalence determination process and verification audits with EU officials from the Food and Veterinary Office (Grange, Ireland) to gain a better understanding of its programs and oversight controls for seafood imports. We included the EU program in the scope of our work because the EU is the largest importer of seafood in the world.

To review the extent to which FDA and FSIS collaborate and coordinate in imported seafood and catfish safety programs, we reviewed a 1984 memorandum of understanding (MOU) signed by FDA, FSIS, and the Environmental Protection Agency that was developed to help promote more effective, efficient, and coordinated federal regulatory activities concerning drug residues. We reviewed the general activities of the Interagency Residue Control Group and Surveillance Advisory Team, which constitute the primary vehicles through which these agencies coordinate their regulatory activities concerning the presence of drug residues. We reviewed the April 2014 MOU between FDA and FSIS that was developed, in part, to improve interagency cooperation on food safety and fraud prevention with regard to the transfer of catfish inspection from FDA to FSIS. We reviewed steps FDA and FSIS have taken to collaborate on transferring responsibility for the oversight of catfish. We reviewed general information on the multi-residue methods (MRM) FDA is developing to test for drug residues in seafood and the MRM FSIS developed for testing catfish for drug residues. We reviewed a March 2010 USDA Office of Inspector General report on how the Interagency Residue Control Group and Surveillance Advisory Team were established and currently functioning. We compared their coordination activities to key practices that can help enhance and sustain federal agencies' collaboration that we previously identified, particularly the practice of identifying and addressing needs by leveraging resources when responsibilities cut across more than one federal agency.[53]

---

[53] GAO, *Results-Oriented Government: Practices That Can Help Enhance and Sustain Collaboration among Federal Agencies*, GAO-06-15 (Washington, D.C.: Oct. 21, 2005).

We met with FDA and FSIS officials to discuss information for all objectives, including obtaining their views on the pros and cons of opportunities to strengthen their programs regarding unsafe drug residues in imported seafood. We reviewed past GAO reports relevant to this topic.[54]

We interviewed CBP officials in headquarters and at selected U.S. ports of entry to gain a better understanding of the agencies' programs. We interviewed CBP officials at the Port of New York in Newark, New Jersey—the largest port of entry for seafood products in the United States—and toured CBP's facilities to observe its examination of seafood. We also interviewed FSIS officials at a reinspection facility in Newark and observed the reinspection process. In addition, we visited and interviewed CBP officials at the Port of Long Beach in Long Beach, California—the second largest port of entry for seafood in the United States. Further, we interviewed CBP officials at the Otay Mesa Land Crossing and Cargo Facility in Otay Mesa, California, to learn about CBP's activities related to ensuring the safety of seafood imports and CBP's interaction with FDA and FSIS, and observed CBP's review process for imported seafood. The Otay Mesa crossing is in proximity to the Long Beach Port. We selected the largest and second largest ports of entry for seafood into the United States and selected a land port that is in close proximity to one of the seaports. We also interviewed FDA officials at the agency's Southwest Import District's Resident Post at Otay Mesa, California, and at the Los Angeles District Office and Pacific Regional Laboratory in Irvine, California, to discuss FDA's testing of seafood imports for drug residues with laboratory officials. In addition, we visited an FSIS import inspection establishment in Vernon, California—close to the Long Beach port—to learn about the measures FSIS uses to ensure that imported catfish do not

---

[54] GAO, *Seafood Safety: Status of Issues Related to Catfish Inspection*, GAO-17-289T (Washington, D.C.: Dec. 7, 2016); *Food Safety: FDA Can Better Oversee Food Imports by Assessing and Leveraging Other Countries' Oversight Resources,* GAO-12-933 (Washington, D.C.: Sept. 28, 2012); *Seafood Safety: Responsibility for Inspecting Catfish Should Not Be Assigned to USDA,* GAO-12-411 (Washington, D.C.: May 10, 2012); *Seafood Safety: FDA Needs to Improve Oversight of Imported Seafood and Better Leverage Limited Resources*, GAO-11-286 (Washington, D.C.: Apr 14, 2011); GAO-06-15. Also see Related GAO Products at the end of the report.

have unsafe drug residues. We also interviewed officials from the National Marine Fisheries Service in the Department of Commerce to gather information on their imported seafood inspection services performed on a fee-for-service basis for private companies, including any data on sampling of imported seafood for drug residues and associated results.

For informational purposes, we spoke with representatives from one large and one medium-sized catfish farm, Tackett Fish Farms and Pentecost Brothers, respectively; one large catfish processor, Heartland Catfish Company; a large feed mill that produces medicated feed, Fishbelt Feed, Inc.; and the National Warmwater Aquaculture Center at the Mississippi State University Extension Service. To gain stakeholders' perspectives on FDA's and FSIS's efforts to address the safety of seafood imports, we also spoke with representatives from the National Fisheries Institute. According to its website, the institute focuses on education about seafood safety and other issues, and includes more than 200 member companies including seafood processors, importers, and exporters. We also spoke with a representative of the Consumer Federation of America. According to its website, the Consumer Federation of America is an association of non-profit consumer organizations that was established to advance the consumer interest through research, advocacy, and education. Today, nearly 300 groups participate in the federation and govern it through representatives on the organization's Board of Directors.

We conducted this performance audit from January 2016 to September 2017 in accordance with generally accepted government auditing standards. Those standards require that we plan and perform the audit to obtain sufficient, appropriate evidence to provide a reasonable basis for our findings and conclusions based on our audit objectives. We believe that the evidence obtained provides a reasonable basis for our findings and conclusions based on our audit objectives.

APPENDIX II: FDA'S DRUG RESIDUE TESTING FOR IMPORTED SEAFOOD

Table 1 shows the number of seafood samples that the Food and Drug Administration tested for drugs and the number and percentage of those samples that contained unsafe drug residues, as well as the volume of seafood imported, for fiscal years 2010 through 2015.

Table 2. Numbers of Imported Seafood Samples Tested by the Food and Drug Administration (FDA) and Percentage Found to Contain Unsafe Drug Residues, and Volume of Seafood Imported, Fiscal Years 2010 through 2015

Seafood type	Fiscal year	Volume imported (in millions of pounds)	Samples tested for drugs	Samples containing unsafe drug residues	Unsafe samples as a percentage of total samples
Catfish	2010	137	88	2	2.3
	2011	203	74	1	1.4
	2012	237	134	7	5.2
	2013	259	99	0	0
	2014	237	67	0	0
	2015	250	33	3	9.1
Salmon	2010	515	105	1	1.0
	2011	529	93	0	0
	2012	618	90	0	0
	2013	655	87	1	1.2
	2014	695	68	0	0
	2015	758	86	0	0
Shrimp	2010	1,234	388	9	2.3
	2011	1,269	415	19	4.6
	2012	1,177	336	12	3.6
	2013	1,113	421	18	4.3
	2014	1,251	396	30	7.6
	2015	1,292	550	67	12.2
Tilapia	2010	474	298	14	4.7
	2011	432	316	6	1.9
	2012	503	274	3	1.1
	2013	504	283	4	1.4
	2014	507	200	1	.5
	2015	495	258	28	10.9

Table 2. (Continued)

Seafood type	Fiscal year	Volume imported (in millions of pounds)	Samples tested for drugs	Samples containing unsafe drug residues	Unsafe samples as a percentage of total samples
Other	2010	3,169	129	17	13.2
	2011	3,086	172	26	15.1
	2012	3,030	168	6	3,6
	2013	3,097	155	11	7.1
	2014	3,106	116	11	9.5
	2015	3,142	213	14	6.6
Total	**2010**	**5,529**	**1,008**	**43**	**4.3**
	2011	**5,519**	**1,070**	**52**	**4.9**
	2012	**5,564**	**1,002**	**28**	**2.8**
	2013	**5,627**	**1,045**	**34**	**3.3**
	2014	**5,798**	**847**	**42**	**5.0**
	2015	**5,937**	**1,140**	**112**	**9.8**
	2010-2015	**33,974**	**6,112**	**311**	**5.1**

Source: GAO analysis of Food and Drug Administration, Department of Commerce, and U.S. Customs and Border Protection data. I GAO-17-443

Note: The number of samples tested by FDA and the associated results are not intended to be statistically valid for projection purposes. Instead, FDA uses a risk-based computerized tool for admissibility screening and determining which products to examine and test based on various factors, such as past violations history and country of origin. In addition, the sample sizes that are particularly small may convey a level of precision that can be misleading because they can greatly change with minor changes in the data.

APPENDIX III: COMMENTS FROM THE DEPARTMENT OF HEALTH AND HUMAN SERVICES

DEPARTMENT OF HEALTH & HUMAN SERVICES

OFFICE OF THE SECRETARY

Assistant Secretary for Legislation
Washington, DC 20201

AUG 3 0 2017

Steve Morris
Director, Natural Resources and Environment
U.S. Government Accountability Office
441 G Street NW
Washington, DC 20548

Dear Mr. Morris:

Attached are comments on the U.S. Government Accountability Office's (GAO) report entitled, *"Imported Seafood Safety: FDA and USDA Could Strengthen Efforts to Prevent Unsafe Drug Residue"* (GAO-17-443).

The Department appreciates the opportunity to review this report prior to publication.

Sincerely,

Barbara Pisaro Clark

Barbara Pisaro Clark
Acting Assistant Secretary for Legislation

GENERAL COMMENTS OF THE FOOD AND DRUG ADMINISTRATION (FDA) ON THE GOVERNMENT ACCOUNTABILITY OFFICE'S DRAFT REPORT ENTITLED: IMPORTED SEAFOOD SAFETY: FDA AND USDA COULD STRENGTHEN EFFORTS TO PREVENT UNSAFE DRUG RESIDUES (GAO-17-443)

The U.S. Department of Health and Human Services (HHS) appreciates the opportunity from the Government Accountability Office (GAO) to review and comment on this draft report.

GAO Recommendation
The Commissioner of the Food and Drug Administration (FDA) should pursue formal agreements with countries exporting seafood to the United States that commit these countries to test for drugs of concern to FDA and the corresponding maximum residual levels (MRLs) that FDA established for these drugs.

HHS Response
HHS partially concurs with GAO's recommendation.

While FDA has not received any requests to establish this type of arrangement, we concur that the Agency can explore pursuing formal arrangements with countries exporting seafood to the United States that commit these countries to test for drugs of concern to FDA and the corresponding MRLs established for these drugs. Factors outside the agency's control that could limit robust implementation of this recommendation include the country's ability and readiness to comply with the requirements necessary to have a successful arrangement. Applicable test methods, analytical capacity and adequate government oversight would be among essential criteria. FDA will continue to use its existing tools to ensure the safety of seafood imported into the United States and, in the event an arrangement is signed, may consider appropriate modifications.

GAO Recommendation
The Commissioner of FDA should coordinate and communicate with FSIS in developing drug residue testing methods and corresponding maximum residue levels for imported seafood that may also be applicable to imported catfish.

HHS Response
HHS concurs with GAO's recommendation.

FDA has a process in place to notify the U.S. Department of Agriculture (USDA) Food Safety and Inspection Service (FSIS) of new tolerances and changes in tolerances for FSIS–regulated products. FDA will extend this process to notify FSIS of concentrations of specific unapproved drugs in catfish over which FDA has public health concerns.

In addition, FDA has contacted FSIS about re-joining the quarterly meetings FDA holds on aquaculture method prioritization and development. The primary FSIS personnel who should attend the Aquaculture method prioritization and development meetings are those involved in laboratory residue analysis. At the annual Surveillance Advisory Team (SAT) review of the FSIS United States National Residue Program (NRP) on June 15, 2017, FSIS and FDA agreed to work on better communication about the FSIS catfish program at the monthly Interagency Residue Control

Group (IRCG) meetings. The SAT is an interagency committee comprised of representatives from FSIS, FDA, the Environmental Protection Agency (EPA), USDA Agricultural Marketing Service, USDA Agricultural Research Service, and the Centers for Disease Control and Prevention (CDC). It consists of experts in veterinary medicine, toxicology, chemistry, and public health who provide professional advice, as well as information on veterinary drug and pesticide use in animal husbandry. The purpose of the SAT is to enhance communication, which includes obtaining and evaluating relevant toxicity and exposure information for each compound that supports the NRP. The IRCG is a large meeting with input from many parts of FSIS, other parts of USDA, FDA, EPA and CDC.

APPENDIX IV: COMMENTS FROM THE U.S. DEPARTMENT OF AGRICULTURE

USDA

United States Department of Agriculture

Food Safety and Inspection Service

1400 Independence Avenue, SW, Washington, D.C. 20250

AUG 24 2017

Steve D. Morris
Director
Natural Resources and Environment
United States Government Accountability Office
441 G Street, N.W.
Washington, DC

Dear Mr. Morris,

The United States Department of Agriculture (USDA) appreciates the opportunity to review the U.S. Government Accountability Office's (GAO) draft report entitled *Imported Seafood Safety: FDA and USDA Could Strengthen Efforts to Prevent Unsafe Drug Residues* (GAO-17-443).

General Comments

The draft report contains a few either misleading or inaccurate statements that GAO uses to support recommendations directed at USDA. We address these below and provide our planned corrective actions for each of the recommendations for executive action that are directed at USDA.

The draft report (page 25) states the following: "FSIS has not made farm visits a routine part of initial equivalence determinations and verification on-site audits..." This statement is somewhat misleading. FSIS has not made any initial equivalence determinations for foreign Siluriformes fish inspection systems. Therefore, FSIS has not yet had the opportunity to conduct on-site verification audits of foreign Siluriformes fish inspection programs for initial equivalence determinations.

The draft report (page 32) states the following: "However, while FSIS will test domestic catfish for at least 61 drugs using its [Multi-Residue Method] MRM and other methods, it is not clear how many drugs or which drugs other countries will test for in catfish exported to the United States, because FSIS will not have such specific requirements for a foreign country's residue monitoring plan after the transition period." This statement appears to imply that FSIS will not know how many drugs or which drugs other countries will test for in Siluriformes fish exported to the United States. This is incorrect, as these countries are required to provide this information to FSIS as part of the Self-Reporting Tool (SRT), which is a requirement for an initial or ongoing equivalence determination.

The draft report (page 35) states the following: "Nevertheless, during the course of our work, we found examples where FDA and FSIS officials, including program directors and laboratory officials, said that they were unaware of the other agency's development and use of more efficient methods to detect drug residues in seafood, including catfish." We believe the characterization of your findings is overstated.

While your audit work may have found examples where officials were unaware of a specific activity, ARS, FSIS, and FDA regularly exchange information related to analytical methods for identifying and quantifying residues of drugs, pesticides, and environmental contaminants.

Finally, it is more accurate to use the term Siluriformes fish throughout this report, as it is the order of fish under FSIS jurisdiction. While the explanation that you provided in the official draft report is correct, the purpose of the amendment (Sec. 12106) in the Agricultural Act of 2014 was to clarify that fish of the order Siluriformes are an amenable species, subject to the appropriate provisions of the Federal Meat Inspection Act. By using the term "catfish" throughout the report, it appears to invalidate the clarification in the 2014 amendment, which GAO may not have intended.

USDA Responses to GAO Recommendations for Executive Action

GAO Recommendation:
The Administrator of FSIS, when doing an on-site audit in another country for an equivalence determination, should visit at least a sample of farms whose catfish are exported to the United States to determine the conditions under which the catfish are being raised, including the drugs being used.

USDA Response:
FSIS partially agrees with this recommendation. The Agricultural Act of 2014 (2014 Farm Bill) requires FSIS to take into account the conditions under which Siluriformes fish are raised. FSIS will perform on-site verification audits as part of its initial equivalence process to ensure that the foreign country's on-farm controls for producing Siluriformes fish and fish products are implemented as documented. This audit may include farm visits as part of the onsite audits. For example, if establishments in a country have produced Siluriformes fish products that the Central Competent Authority (CCA) or FSIS has found positives of chemicals or pesticides, FSIS would include farm visits as part of the audit. In other situations, FSIS might not visit the farm. However, during the audit, the auditor would verify that the CCA has sufficient on-farm controls based on records reviews at the CCA offices and observations of activities and records at establishments and laboratories.

GAO Recommendation:
The Administrator of FSIS should require as part of an equivalence determination that countries exporting catfish to the United States include in their residue monitoring plans the drugs of concern to FSIS and the corresponding maximum residue levels.

USDA Response:
FSIS' current policy already addresses this recommendation. For a foreign country to be eligible to export meat (which may include Siluriformes fish), poultry, or egg products to the United States, it must show that its inspection system for meat, poultry, or egg products is equivalent to that of the United States. This means that the country must demonstrate that it either employs the same measures as the United States, and many foreign countries take this approach, or that the measures that it employs, while different from those of the United States, provide an

equivalent level of public health protection as those of the United States. To be equivalent, FSIS requires countries to routinely assess their chemical residue plans to identify potential chemical residue concerns unique to their country and develop a plan that addresses those chemical concerns. Accordingly, FSIS recognizes there to be certain differences in foreign countries' chemical residue programs because of differences in how countries address chemical residue concerns. Countries must demonstrate that their measures are as effective as those that FSIS takes to address a food safety issue. FSIS must find that a foreign country is employing measures that ensure that there are no unsafe residues of the drugs in product that the country wishes to ship to the United States, or that country is not eligible to export to the United States.

Foreign countries that are currently shipping Siluriformes fish or Siluriformes fish product to the United States must submit in their SRTs, by September 1, 2017 (the end of the transitional period), documentation demonstrating that their system for controlling chemical residues provides an equivalent level of protection to that of the United States. No country has been determined yet to be equivalent for Siluriformes fish or Siluriformes fish products. For initial equivalence, the country must provide information about its residue sampling program to support that the program is equivalent. The foreign country is required to describe the design of its chemical residue control program, its basis (e.g., whether it is risk based or statistically based), and the government's role in oversight of the development and implementation of the program. The submission must also include information about the overall design, analytical methods, how the government decides which chemicals to sample and analyze for, and how they react to violative findings. The country must also provide FSIS with a sampling plan that describes the frequency of sampling, number of samples collected, and analyses performed for each sampling location and product.

FSIS conducts an in-depth review of the foreign country's SRT submission, including information submitted about its chemical residue control program to determine whether it is equivalent. Fish exported to the United States may contain residues only within the tolerances set by FDA and EPA at 21 CFR 556 and 40 CFR 180, respectively. FSIS would question the design of a program if compounds are detected at FSIS' point of entry testing that are not addressed by the country's residue plan, or if FSIS has other reasons to believe that the compounds are in active use in the exporting country and the country is not testing for them. If a foreign country does not have a residue program that is equivalent to FSIS' program, it does not have an equivalent inspection system, and thus would not be determined eligible to continue exporting Siluriformes fish and fish products to the United States.

To maintain ongoing equivalence after an initial equivalence determination, the foreign country must submit the following: a complete list of chemical residues that will be sampled and analyzed in the upcoming year, the analytical methods for each chemical compound, the analytical results from the previous year, and a description of how and what changes are made to its sampling plans one year to the next.

GAO Recommendation:
The Administrator of FSIS should coordinate and communicate with FDA in developing drug residue testing methods and corresponding maximum residue levels for imported catfish that may also be applicable to other imported seafood.

USDA Response:
FSIS partially agrees with this recommendation. FSIS will coordinate and communicate with FDA in developing drug residue methods and corresponding maximum residue levels for imported Siluriformes fish. Extension of either methods or detection levels to other seafood commodities is beyond FSIS' regulatory responsibility. FSIS fully intends to implement the provisions of the Memorandum of Understanding (MOU) with FDA on coordinating on testing methods, and will enhance residue testing coordination and collaboration through the Surveillance Advisory Team (SAT) and Interagency Residue Control Group (IRCG) meetings.

Relatedly, the Agricultural Research Service (ARS) undertakes or coordinates research that addresses FSIS' needs regarding the development and validation of methods for the detection of drug residues, including Siluriformes fish, with one project currently tasked with this effort. Research accomplishments are transferred to FSIS when applicable and technically possible. FDA is also made aware of analytical methods through various means, such as the publication of adopted FSIS methods in the Chemistry Laboratory Guidebook, and at Annual Research Meetings with ARS, FSIS, and FDA.

Again, thank you for the opportunity to review and comment on this draft report.

Sincerely,

Carmen Rottenberg
Acting Deputy Under Secretary
Office of Food Safety

In: Food Safety
Editor: Georges Martin

ISBN: 978-1-53616-762-7

Chapter 4

FOOD SAFETY: USDA SHOULD TAKE FURTHER ACTION TO REDUCE PATHOGENS IN MEAT AND POULTRY PRODUCTS*

United States Government Accountability Office

ABBREVIATIONS

AMS	Agricultural Marketing Service
APHIS	Animal and Plant Health Inspection Service
CDC	Centers for Disease Control and Prevention
FDA	Food and Drug Administration
FSIS	Food Safety and Inspection Service
HACCP	Hazard Analysis and Critical Control Point
USDA	U.S. Department of Agriculture.

* This is an edited, reformatted and augmented version of United States Government Accountability Office, Report to Congressional Requesters, Publication No. GAO-18-272, dated March 2018.

WHY GAO DID THIS STUDY

The U.S. food supply is generally considered safe, but the Centers for Disease Control and Prevention (CDC) estimate that *Salmonella* and *Campylobacter* in food cause about 2 million human illnesses per year in the United States. In 2014, GAO identified challenges USDA faced in reducing pathogens in poultry products, including standards that were outdated or nonexistent and limited control over factors that affect pathogen contamination outside of meat and poultry slaughter and processing plants, such as practices on the farm.

GAO was asked to review USDA's approach to reducing pathogens in meat and poultry products. This chapter examines (1) the extent to which USDA has developed standards for meat and poultry products and (2) any additional steps USDA has taken to address challenges GAO identified in 2014. GAO reviewed relevant regulations, documents, and data and interviewed officials from USDA and CDC, as well as 17 stakeholders representing industry, consumer groups, and researchers selected based on their knowledge of USDA's meat and poultry slaughter inspections and food safety.

WHAT GAO RECOMMENDS

GAO is making three recommendations, including that USDA document its process for deciding which products to consider for new standards and that it include information on the effectiveness of on-farm practices in its guidelines for *Salmonella* control in hogs. USDA agreed with GAO's recommendations and described actions it will take to implement them.

WHAT GAO FOUND

To help ensure the safety of our nation's food supply, the U.S. Department of Agriculture (USDA) has developed standards limiting the amount of *Salmonella* and *Campylobacter*—pathogens that can cause foodborne illness in humans— permitted in certain meat (beef and pork) and poultry (chicken and turkey) products, such as ground beef, pork carcasses, and chicken breasts. However, the agency has not developed standards for other products that are widely available, such as turkey breasts and pork chops. Further, its process for deciding which products to consider for new standards is unclear because it is not fully documented, which is not consistent with federal standards for internal control. For example, USDA has informed stakeholders that it will take into account factors including consumption and illness data, but the agency has not documented this process going forward. Previously, USDA had developed new standards after widespread outbreaks indicated the need. For example, in 2016, USDA concluded that new standards were needed for certain poultry products to reduce *Salmonella* after reviewing outbreaks from these products in 2011, 2013, and 2015—outbreaks in which 794 people were sickened and 1 died. By documenting the agency's process for deciding which products to consider for new standards, USDA could better ensure that such decisions will be risk-based.

USDA is taking steps to address challenges GAO identified in 2014 for reducing pathogens in poultry products, but these challenges are ongoing and could affect USDA's ability to reduce pathogens in meat as well. For example, one challenge GAO identified is that the level of pathogens in poultry products can be affected by practices on farms where poultry are raised. GAO recommended in 2014 that to help overcome this challenge, USDA guidelines on practices for controlling *Salmonella* and *Campylobacter* on farms include information on the effectiveness of each of the practices, consistent with a recommendation from a USDA advisory committee. Since GAO's 2014 report, USDA drafted revised guidelines to include information on the effectiveness of on-farm practices for controlling pathogens in poultry and beef cattle, in 2015 and 2017, respectively. However, USDA's draft guidelines for controlling *Salmonella* in hogs do not contain such information. By including such information as it finalizes its draft guidelines, USDA could better inform industry of the potential benefits of adopting on-farm practices included in the guidelines and encourage implementation of such practices.

March 19, 2018

The Honorable Kirsten E. Gillibrand
Ranking Member
Subcommittee on Livestock, Marketing and Agriculture Security
Committee on Agriculture, Nutrition and Forestry
United States Senate

The Honorable Dianne Feinstein
United States Senate

The Honorable Richard J. Durbin
United States Senate

The U.S. food supply is generally considered safe, but foodborne illness remains a common, costly, yet largely preventable public health problem. The Centers for Disease Control and Prevention (CDC) estimated in 2011—its most recent estimate—that each year, one in six people in the United States gets sick from eating contaminated food. According to CDC, while the source of foodborne outbreaks is often unknown,[1] certain pathogens in food cause more than 9 million human illnesses per year in the United States, 2 million of which are caused by *Salmonella* and *Campylobacter*.[2] For example, according to a CDC outbreak investigation

---

[1] According to CDC's website, when two or more people get the same illness from the same contaminated food or drink, the event is called a foodborne outbreak.

[2] Pathogens are disease-causing organisms and include bacteria such as *Salmonella* and *Campylobacter*.

report, in 2015, 192 people in five states became ill from eating pork contaminated with *Salmonella*.[3]

The U.S. Department of Agriculture (USDA) is responsible for ensuring the safety and wholesomeness of meat and poultry products that enter commerce, as provided by the Federal Meat Inspection Act and the Poultry Products Inspection Act.[4] Accordingly, USDA's Food Safety and Inspection Service (FSIS) inspects and regulates the production of domestic meat and poultry products sold for human consumption. For meat and poultry processing and slaughter plants (hereafter referred to as plants) under its jurisdiction, FSIS sets pathogen reduction performance standards (hereafter referred to as pathogen standards) limiting the allowable levels of *Salmonella* or *Campylobacter*.[5] FSIS personnel perform inspection activities at nearly 6,500 plants nationwide, including testing samples of meat and poultry products to assess compliance with the pathogen standards, which vary by product. For fiscal year 2016, Congress appropriated more than $1 billion to FSIS for its activities, including its food safety inspection program, through the annual appropriations process.

In September 2014, we found that USDA had taken a number of actions to reduce the levels of *Salmonella* and *Campylobacter* in poultry (chicken and turkey) products but that the agency needed to do more to assess the effects of these actions on the incidence of foodborne illnesses.[6] As a result, we recommended that USDA develop performance measures to monitor whether its activities to reduce the levels of *Salmonella* and *Campylobacter* in domestic poultry products are meeting agency goals. USDA agreed with and implemented this recommendation. We also

---

[3] Centers for Disease Control and Prevention, *2015 Outbreaks: Multistate Outbreak of Multidrug-Resistant Salmonella I 4, [5],12:i:- and Salmonella Infantis Infections Linked to Pork* (Final Update) (Atlanta, GA: December 2, 2015).

[4] 21 U.S.C. §§ 601-683 and 21 U.S.C. §§ 451-472.

[5] FSIS pathogen standards apply to meat and poultry products that are raw or not fully cooked. When USDA purchases raw ground beef for certain programs, such as the National School Lunch Program, it does not allow for any *Salmonella* contamination. For additional information, see appendix I.

[6] GAO, *Food Safety: USDA Needs to Strengthen Its Approach to Protecting Human Health from Pathogens in Poultry Products*, GAO-14-744 (Washington, D.C.: Sept. 30, 2014). Additional GAO reports in the area of food safety are listed in "Related GAO Products" at the end of this chapter.

identified several challenges that could hinder USDA's ability to reduce levels of pathogens in poultry products; in particular, we found that pathogen standards for poultry products were outdated or nonexistent. Other challenges we identified included limited USDA control over factors that affect pathogen contamination outside of plants, plants not designating pathogens as hazards, and the complex nature of pathogens, among others. We recommended that FSIS include in its guidelines information on the effectiveness of on-farm practices to reduce the level of pathogens in live poultry. USDA agreed with and is taking steps to implement this recommendation.

However, according to CDC data, consumption of poultry products continues to be associated with foodborne illness in the United States, as does the consumption of meat (such as beef and pork).[7] Further, according to a 2016 CDC report, the United States is not on track to reach its public health goals for reducing foodborne illnesses caused by *Salmonella* and *Campylobacter*.[8]

You asked us to review USDA's approach to reducing the level of pathogens in domestic meat and poultry products, including its use of pathogen standards. This chapter examines (1) the extent to which USDA has developed pathogen standards for meat and poultry products and (2) any additional steps that USDA has taken to address challenges we identified in 2014 in reducing the level of pathogens.

To examine the extent to which USDA has developed pathogen standards for meat and poultry products, we reviewed our prior findings and recommendations from September 2014 on pathogen standards for poultry products.[9] We also reviewed relevant laws and regulations, FSIS strategic plans covering the period from 2011 through 2021, and FSIS annual performance plans from 2014 through 2018. We reviewed relevant *Federal Register* notices on specific pathogen standards for meat and poultry from 1996, when FSIS first established the standards, through

---

[7] Centers for Disease Control and Prevention, *Surveillance for Foodborne Disease Outbreaks, United States, 2015: Annual Report* (Atlanta, GA: 2017).

[8] Centers for Disease Control and Prevention, *Healthy People 2020 Midcourse Review* (Atlanta, GA: 2016).

[9] GAO-14-744.

2016, when the agency set or revised the most recent pathogen standards.[10] We identified relevant performance goals and measures in FSIS annual performance plans from 2012 through 2017. We reviewed FSIS reports on pathogens in meat and poultry products and annual sampling plans—estimates of the type and number of products the agency will test for certain pathogens—to understand the range of meat and poultry products FSIS monitors for pathogens.[11] In addition, we interviewed current and former FSIS officials on the history of the 1996 Pathogen Reduction; Hazard Analysis and Critical Control Point (HACCP) regulations,[12] related pathogen standards, and the process and criteria for developing new pathogen standards and revising existing ones. We reviewed documents and interviewed staff from USDA's Agricultural Marketing Service—which purchases commodities, including meat and poultry products, for federal nutrition assistance programs such as the National School Lunch Program—to determine any additional steps USDA takes to prevent pathogens in meat and poultry products used for school meals.[13] We reviewed data FSIS collected on compliance rates for meat and poultry plants from 2014 through 2017 and data on foodborne illness rates from CDC from 1998 through 2016, the most recent data available. To assess the reliability of the data, we interviewed CDC officials about the steps they took to ensure the reliability of the data and determined that these data were sufficiently reliable for viewing trends in compliance rates and trends in foodborne illness rates.

---

[10] 61 Fed. Reg. 38,806 (July 25, 1996); 81 Fed. Reg. 7285 (Feb. 11, 2016).

[11] FSIS's annual sampling plan describes the major activities related to microbiological sampling and the agency's overall strategy for directing its sampling resources.

[12] 61 Fed. Reg. 38806 (July 25, 1996). Under the HACCP approach, industry—rather than federal inspectors—is responsible for (1) identifying food safety hazards that are reasonably likely to occur and (2) establishing controls that prevent or reduce these hazards. Plants are required to consider hazards that can occur before, during, and after entry into the establishment. As part of the HACCP approach, plants must develop plans that identify the point (known as the critical control point) where they will take steps to prevent, eliminate, or reduce each hazard identified. Plants must also conduct activities to verify the plans are being effectively implemented.

[13] Commodities include foods procured by USDA and provided to states at no charge for schools to serve in school meal programs. School meal programs include the National School Lunch Program, which, in addition to providing lunches, provides after-school snacks; the School Breakfast Program; and Special Milk Program. For additional information, see appendix I.

We identified an initial group of stakeholders from our prior work, specifically from those we interviewed in our 2013 report on USDA's changes to poultry and hog inspections and our 2014 report on poultry pathogens.[14] In addition, we asked these groups for recommendations on other stakeholders we should consider contacting and expanded the list, as needed. We selected these stakeholders because they are knowledgeable about FSIS's food safety programs and provide a range of views on the topic. In the end, we identified a nongeneralizable sample of 17 stakeholders: 7 representatives from industry, 4 representatives from consumer advocacy groups, 4 food safety researchers, and 2 former federal food safety experts. Views from those we selected based on their knowledge cannot be generalized to all stakeholders who have knowledge about FSIS's food safety programs (i.e., those we did not interview), but they provide illustrative examples. We obtained information from agency documentation and interviews with FSIS officials regarding the process for developing new pathogen standards and compared this process with federal standards for internal control.[15] We also obtained information from FSIS documentation and interviews with agency officials on the agency's plans to review or revise pathogen standards. We compared these plans with the Project Management Institute's standards and leading practices for project management.[16]

To examine any additional steps that USDA has taken to address the challenges we identified in 2014 that it faces in reducing the level of pathogens in meat and poultry, we reviewed agency documentation on the steps it has taken to address these challenges since 2014, including documentation on relevant laws and regulations; *Federal Register* notices; FSIS's 2017-2021 strategic plan, annual performance plans and related

---

[14] GAO, *Food Safety: More Disclosure and Data Needed to Clarify Impact of Changes to Poultry and Hog Inspections*, GAO-13-775 (Washington, D.C.: Aug. 22, 2013); GAO-14-744.

[15] GAO, *Standards for Internal Control in the Federal Government,* GAO-14-704G (Washington, D.C.: September 2014).

[16] The Project Management Institute is a not-for-profit association that provides global standards for, among other things, project and program management. These standards are utilized worldwide and provide guidance on how to manage various aspects of projects, programs, and portfolios. See Project Management Institute, Inc., *The Standard for Program Management®*, Third Edition (2013).

performance reports from 2015 through 2017, and 2016 Establishment-Specific Data Release Strategic Plan;[17] and USDA and FSIS websites. We reviewed FSIS guidance regarding on-farm practices for reducing the level of pathogens in live animals before they enter plants and compared this information with recommendations on such practices made in 2011 by the USDA National Advisory Committee on Meat and Poultry Inspection. We also reviewed reports from the Interagency Food Safety Analytics Collaboration,[18] USDA's Office of Inspector General, and the National Academy of Sciences.

We obtained testing data from FSIS for the period from May 2016 through August 2017, the most recent data available, to determine the number of plants that did not meet the chicken carcass *Salmonella* and *Campylobacter* standards. To assess the reliability of FSIS's data, we interviewed FSIS officials about the steps they took to ensure the reliability of the data and determined that the data were sufficiently reliable for analyzing the number of plants that did not meet the pathogen standards during the reporting period. We reviewed CDC data on foodborne illness outbreaks from beef and pork from 2006 through 2016 and obtained recall documents from FSIS on plants that produced products involved in those outbreaks. We interviewed FSIS officials about any challenges the agency faces in reducing pathogens in meat and poultry and steps the agency has taken since 2014 to address previously identified challenges. Additionally, we interviewed FSIS and CDC officials on steps the agencies have taken to improve the methods used to estimate the burden of foodborne illness in the United States and the use of emerging technologies to detect the pathogen strains most harmful to human health. Further, we interviewed

---

[17] FSIS's 2016 establishment-specific data release strategic plan is separate from the agency's strategic plan. The document provides a framework under which FSIS plans to responsibly and effectively release plant-level data collected by the agency. Among other things, it describes FSIS's data collection activities, data release procedures, and a prioritized list of data for public release.

[18] To enhance the safety of our food, CDC, USDA, and FSIS teamed up in 2011 to create the Interagency Food Safety Analytics Collaboration. According to the interagency collaboration's website, its goal is to improve the coordination of federal food safety analytic efforts and address cross-cutting priorities for food safety data collection, analysis, and use. Projects and studies by the interagency collaboration aim to identify foods that are important sources of illnesses.

the stakeholder groups we identified for our first objective about USDA's efforts to address the challenges we identified in 2014.

We conducted this performance audit from June 2016 to January 2018 in accordance with generally accepted government auditing standards. Those standards require that we plan and perform the audit to obtain sufficient, appropriate evidence to provide a reasonable basis for our findings and conclusions based on our audit objectives. We believe that the evidence obtained provides a reasonable basis for our findings and conclusions based on our audit objectives.

BACKGROUND

Contact with infected animals or consumption of contaminated water and food—including produce, meat, poultry, and processed products—can cause foodborne illness. Many different pathogens can contaminate food, including harmful bacteria such as *Salmonella* and *Campylobacter*. CDC reported that in 2015 there were 902 foodborne disease outbreaks reported in the United States that resulted in 15,202 illnesses, 950 hospitalizations, 15 deaths, and 20 food product recalls.[19] According to CDC, fish, chicken, and pork were the most common single food categories implicated in these outbreaks. More recently, in 2016, there were 233 foodborne illnesses from 10 outbreaks linked to beef, 426 foodborne illnesses from 17 outbreaks linked to pork, and 417 foodborne illnesses from 20 outbreaks linked to poultry, according to CDC's National Outbreak Reporting System (see Figure 1).[20] Common symptoms of foodborne diseases include nausea, vomiting, stomach cramps, and diarrhea. Symptoms can sometimes be severe, and some foodborne illnesses can be life-threatening. Although

---

[19] Centers for Disease Control and Prevention, *Surveillance for Foodborne Disease Outbreaks United States, 2015: Annual Report* (Atlanta, GA: 2017).

[20] These data are from CDC's National Outbreak Reporting System, a web-based platform that uses a standard data collection form. An outbreak reported to the system is defined as the occurrence of two or more cases of a similar illness resulting from a common exposure. Reporting of outbreaks to CDC is voluntary; outbreaks are likely underreported due to limited resources and training in health departments.

anyone can get a foodborne illness, some people are more likely to have one. Those groups include young children, older adults, pregnant women, and people with immune systems weakened from medical conditions, such as diabetes, liver, and kidney disease. Patients receiving chemotherapy or radiation treatment are also more susceptible.

Salmonella

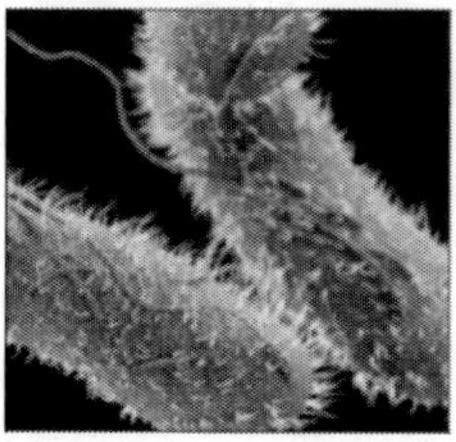

According to the Centers for Disease Control and Prevention (CDC), Salmonella bacteria can cause illness in humans—salmonellosis. There are thousands of serotypes—groups within a single species that share distinctive surface structures—of Salmonella but not all cause illness in people.
CDC estimates that approximately 1 million illnesses occur annually in the United States because of Salmonella in food. Most people infected with Salmonella develop diarrhea, fever, and abdominal cramps within 12 to 72 hours. The illness usually lasts 4 to 7 days, and most individuals recover without treatment. In some cases, however, diarrhea may be so severe that the patient needs hospitalization, according to CDC. The Salmonella infection may also spread from the intestines to the bloodstream, and then to other body sites. In these cases, Salmonella can cause death unless the person is treated promptly with antibiotics. The elderly, infants, and those with impaired immune systems are more likely to have a severe illness.

Source: GAO analysis of information provided by CDC. Image: Courtesy of CDC, James Archer. | GAO-18-272.

We have previously reported that to improve its food safety approach, FSIS moved to an increasingly science-based, data-driven, risk-based approach by adopting the Pathogen Reduction; HACCP regulations in 1996.[21] Under the HACCP approach, each plant is responsible for (1) identifying food safety hazards, such as fecal material, that are reasonably likely to occur and (2) establishing controls that prevent or reduce these hazards in its processes. As part of this approach, plants must develop plans that identify the point (known as the critical control point) where they will take steps to prevent, eliminate, or reduce each hazard identified.

---

[21] GAO-14-744; 61; *Fed. Reg.* 38,806 (July 25, 1996).

FSIS inspectors at slaughter and processing plants routinely check records to verify a plant's compliance with those plans. FSIS inspectors also observe operations at plants as part of their inspection activities.

Campylobacter

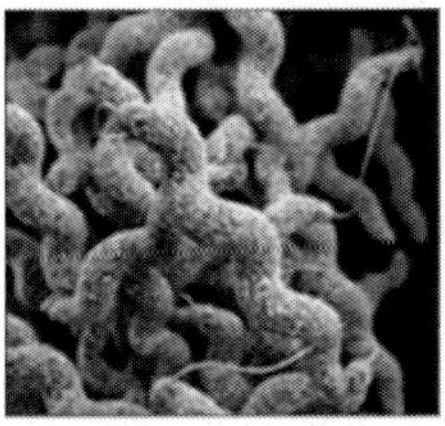

According to the Centers for Disease Control and Prevention (CDC), Campylobacter bacteria can cause illness in humans— campylobacteriosis. In contrast, Campylobacter seems to be well adapted to birds, which carry it without becoming ill. Most campylobacteriosis in humans is caused by one species, Campylobacter jejuni, but other species can also cause human illness. Campylobacter can infect anyone, but campylobacteriosis is more common in males, children younger than 5 years, and people ages 65 years and older. According to CDC, Campylobacter is one of the most common causes of diarrheal illness in the United States; other common symptoms include fever, abdominal cramps, nausea, and vomiting. Some infected people do not have any symptoms. Symptoms usually start within 2 to 5 days after exposure and last about a week. Most cases occur as isolated events, not as part of recognized outbreaks. CDC estimates that Campylobacter in food causes 850,000 illnesses each year in the United States. Campylobacteriosis occurs more frequently in the summer months than in the winter. Although most people who get ill from Campylobacter recover completely, according to a 2011 CDC study on foodborne illnesses acquired in the United States, approximately 76 persons with Campylobacter infections die each year.

Source: GAO analysis of information provided by CDC. Image: Courtesy of CDC, James Archer. | GAO-18-272.

Under the 1996 HACCP regulations, the agency also established *Salmonella* pathogen standards used to assess the effectiveness of plants' controls in reducing levels of pathogens in meat and poultry products. According to the regulations, FSIS selected *Salmonella* for pathogen standards because, among other things, it was the most common bacterial cause of foodborne illness, and they believed that intervention strategies aimed at reducing fecal contamination and other sources of *Salmonella* on raw product should be effective against other pathogens.[22] FSIS has a

[22] According to CDC's website, other foodborne pathogens found in meat and poultry include *Clostridium perfringens, Listeria monocytogenes*, and *Yersinia enterocolitica*.

verification-testing program in which FSIS inspectors at plants collect samples of certain products and test them to determine whether plants meet the pathogen standards. Test results from this program help FSIS inspectors verify that plant HACCP plans are working and identify and assist plants whose process controls may be underperforming. FSIS also requires products to be labeled with instructions for safe handling.

Multistate Outbreak of *Escherichia coli O157:H7* Infections from Hamburgers, 1992-1993

In the early 1990s, a strain of Escherichia coli (E. coli) bacteria linked to hamburger resulted in a deadly foodborne outbreak and led to changes in food regulations. E. coli are bacteria found in the environment, food, and intestines of people and animals. E. coli are a large and diverse group of bacteria. Most strains of E. coli are generally harmless, but others can cause diarrhea, urinary tract infections, respiratory illness and pneumonia, other illnesses, and death. From November 1992 through February 1993, more than 500 laboratory-confirmed infections with E. coli O157:H7 and four associated deaths occurred in four states—Washington, Idaho, California, and Nevada. During the outbreak, contaminated hamburger patties were traced to a fast food restaurant chain and then ultimately to five slaughter plants in the United States and one in Canada as likely sources of carcasses used to produce the contaminated ground beef. No one slaughter plant or farm was identified as the source. In 1994, USDA's Food Safety and Inspection Service declared that any raw ground beef found contaminated with E. coli O157:H7 would be adulterated under the Federal Meat Inspection Act—rendering the meat unlawful to sell in commerce.

Source: GAO analysis of Centers for Disease Control and Prevention and U.S. Department of Agriculture (USDA) documents. Image: Courtesy of USDA. | GAO-18-272.

In contrast to *Salmonella* and *Campylobacter*, which are subject to pathogen standards, FSIS considers certain serotypes of *Escherichia coli* (*E. coli*),[23] another type of disease-causing pathogen, adulterants under the

[23] *E. coli* bacteria live in the intestines of healthy cattle and have a symbiotic relationship with the cattle, an association in which the *E. coli* derives the benefit and the cattle are not harmed. A cow with *E. coli* in its intestinal system typically "sheds" the organism through its feces. As cattle shed *E. coli*, the bacteria can contaminate the hides and then the meat as the cattle are slaughtered. Several strains of *E. coli* are highly pathogenic and capable of causing death when they infect humans. See GAO, *Food Safety: Preslaughter Interventions Could Reduce E. coli in Cattle*, GAO-12-257 (Washington, D.C.: Mar. 9, 2012).

definition of "adulterated" in the Meat Inspection Act and the Poultry Inspection Act.[24]

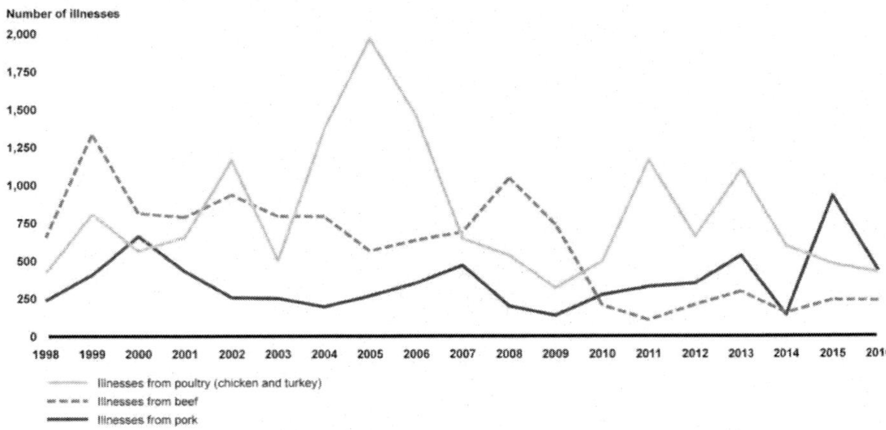

Source: GAO analysis of Centers for Disease Control and Prevertion data. | GAO-18-272.

Notes: These data are from the Centers for Disease Control and Prevention's (CDC) National Outbreak Reporting System, a web-based platform that uses a standard data collection form. An outbreak reported to the system is defined as the occurrence of two or more cases of a similar illness resulting from a common exposure. Reporting of outbreaks to CDC is voluntary; outbreaks are likely underreported because of limited resources and training in health departments. These data include outbreaks in all 50 states (including outbreaks occurring in multiple states), the District of Columbia, and U.S. territories. These data include outbreaks with one or more bacterial, viral, chemical or toxin, parasitic, or unknown causes and suspected and confirmed etiologies. Implicated foods in outbreaks are classified into 1 of 24 single food categories (e.g., beef, pork, chicken, or turkey) if a single contaminated ingredient was identified or if all ingredients belonged to that category.

Figure 1. Foodborne Outbreak-associated Illness Linked to Beef, Pork, and Poultry Products, 1998 through 2016.

The acts define an adulterant in meat and poultry products to include, among other things, "any poisonous or deleterious substance which may render it injurious to health." Meat and poultry products contaminated with

---

[24] 21 U.S.C. §§ 601-683 and 21 U.S.C. §§ 451-472.

any level of adulterants are not permitted to enter commerce—a stricter standard than the pathogen standards, which allow certain levels of contamination. FSIS initially declared *E. coli O157:H7* as an adulterant in ground beef following an outbreak from 1992 to 1993 that involved Jack-in-the-Box hamburgers and, in 2011, declared an additional six non-O157 Shiga-toxin-*forming E. coli* in certain raw beef products as adulterants.

Meat and Poultry in the United States

Beef: According to the U.S. Department of Agriculture (USDA), beef is a highly consumed meat in the United States, averaging 56 pounds per person per year. Beef comes from full-grown cattle that are about 2 years old and weigh about 1,000 pounds. There are at least 50 breeds of beef cattle, but fewer than 10 make up most cattle produced. Veal is meat from a calf (young cattle) that weighs about 150 pounds. Calves that are mainly milk-fed usually are less than 3 months old.
Pork: According to the USDA, the United States is the world's third-largest producer and consumer of pork and pork products. Pork is meat from hogs, or domestic swine, and is from young animals (6 to 7 months old) that weigh from 175 to 240 pounds.

Poultry: According to USDA, consumption of poultry (chicken and turkey) in the United States is higher than beef or pork. Chicken includes broiler-fryer chickens and roaster chickens. Broiler-fryer chickens are young, tender chickens about 7 weeks old that weigh from 2 1/2 to 4 1/2 pounds. Roaster chickens are young chickens from 8 to 12 weeks old with a ready-to-cook carcass weight of 5 pounds or more. Turkey is a large, widely domesticated North American bird. They grow to full maturity in about 4 to 5 months, depending on the desired market weight.

Source: GAO analysis of information provided by USDA. Image: Courtesy of USDA, Lance Cheung. | GAO-18-272.

FSIS coordinates with numerous federal agencies, state agencies, and local entities to help ensure the safety of meat and poultry products from the farm to the consumer (known as the farm-to-table continuum).[25] FSIS coordinates with USDA's Animal and Plant Health Inspection Service

[25] The farm-to-table continuum for food safety includes all facets of the production process: on the farm, animal slaughter in FSIS-regulated plants, food processing within regulated plants, retail or market establishments (e.g., grocery stores), and home environments.

(APHIS) to share information when investigating foodborne illnesses. FSIS also coordinates with the Department of Health and Human Services' Food and Drug Administration (FDA) and with CDC on a number of activities.

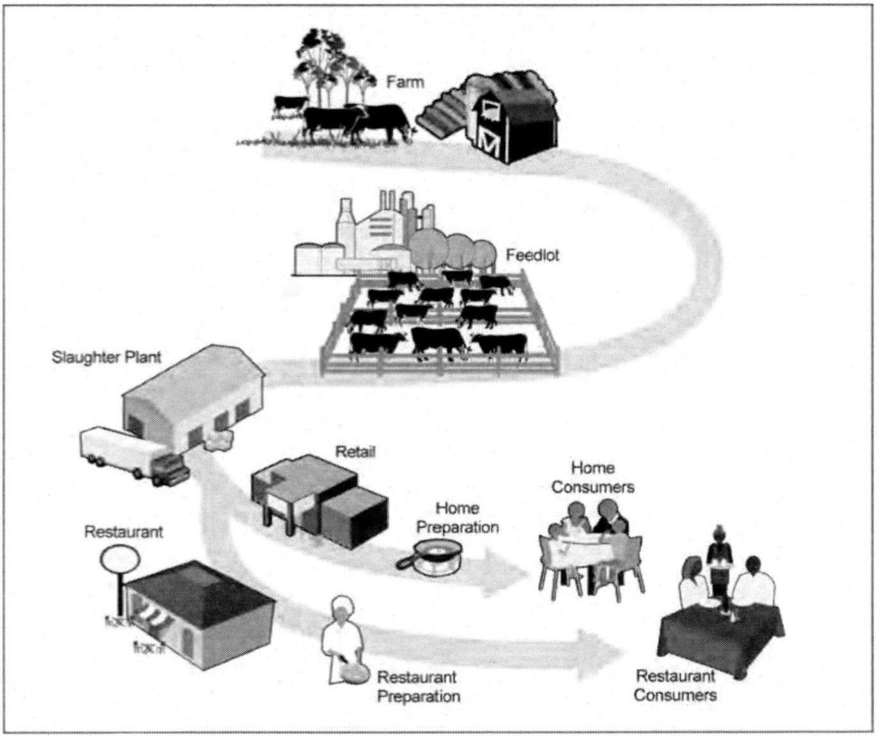

Source: Centers for Disease Control and Prevention and GAO. | GAO-18-272.

Note: According to the U.S. Department of Agriculture, conventional beef production includes placing cattle on feedlots. Other types of cattle are produced through alternate systems. For example, beef from alternate production systems—including natural, organic, and grass- or forage-fed—account for about 3 percent of the U.S. beef market.

Figure 2. Example of Farm-to-Table Continuum for Beef.

For example, FSIS works collaboratively with FDA and CDC through the Interagency Food Safety Analytics Collaboration to, among other things, estimate foodborne illness source attribution. Attribution entails identifying which foods are the most important sources of selected major foodborne illnesses.[26] According to FSIS officials, determining the sources of illness is an important part of identifying opportunities to improve food safety. FSIS also coordinates with CDC and state health departments to respond to foodborne illness outbreaks, including identifying the pathogen, the product, and where the product became contaminated along the farm-to-table continuum (see Figure 2).

Additionally, FSIS, CDC, FDA, representatives from state and local regulatory offices, and stakeholders outside the government developed Healthy People 2020 food safety targets to reduce rates of infection caused by certain foodborne pathogens, including *Salmonella* and *Campylobacter*, by the year 2020. According to FSIS officials, FSIS will continue to co-lead the Healthy People Food Safety Working Group, which includes setting Healthy People 2030 targets for reductions in foodborne illness and monitoring progress.

USDA Has Developed or Revised Standards for Reducing the Level of Pathogens in Certain Products But Not Others

USDA's FSIS has developed or revised pathogen standards for assessing the effectiveness of plants' controls in reducing the level of pathogens in certain meat and poultry products. More specifically, the agency has developed pathogen standards for some beef, pork, chicken, and turkey products but not for other products that are widely available, and its basis for deciding which products to consider for new pathogen standards is unclear. In addition, as of 2011, the agency has revised

---

[26] FSIS is responsible for the safety of meat, poultry, and processed egg products. A provision of the Food, Conservation, and Energy Act of 2008, Pub. L. No. 110-246 § 11016(b) amended the Federal Meat inspection Act to, among other things, subject catfish to the inspection requirements of the Federal Meat Inspection Act. The Agricultural Act of 2014, Pub. L. No. 113-79 §12106(a), among other things, amended the 2008 provision to instead make all fish of the order Siluriformes subject to Federal Meat Inspection Act inspections.. The order *Siluriformes* includes the families of *Ictaluridae*, *Clariidae*, and *Pangasiidae*. FDA is responsible for virtually all other food.

pathogen standards for chicken and turkey products, but standards for other products are outdated, with no time frames for revision.

USDA Has Developed Standards for Some Meat and Poultry Products but Not Others, and Its Basis for Deciding Which Products to Consider for New Standards Is Unclear

FSIS has developed pathogen standards for beef, pork, chicken, and turkey carcasses; specific chicken parts (i.e., breasts, thighs, and legs); and ground beef, chicken, and turkey (see Figure 3). The initial pathogen standards FSIS developed in 1996 were all for *Salmonella* because, among other things, it was the most common bacterial cause of foodborne illness and intervention strategies aimed at reducing *Salmonella* in raw products might be effective against other pathogens, according to agency documents. Subsequently, in 2011, FSIS developed *Campylobacter* standards for chicken and turkey carcasses and in 2016 developed *Salmonella* and *Campylobacter* standards for chicken parts.

FSIS has not developed pathogen standards for other widely available products, such as pork cuts (e.g., pork chops), turkey parts (e.g., turkey breasts), and ground pork. The agency is taking steps that may lead to the development of new pathogen standards for additional products. For example, according to FSIS documents, the agency is collecting information on the presence of *Salmonella* and other pathogens in pork cuts and ground pork, among other pork products.[27] According to FSIS officials, this could lead to the development of new standards.

However, the agency's process for deciding which products to consider for new pathogen standards is unclear because it is not fully documented. In December 2016, the agency documented a part of its process: who will make the decisions about which products to consider.

---

[27] According to FSIS officials, any potential new pathogen standards for pork products may be directed at comminuted pork, which includes ground pork.

Products with *Salmonella* standards	
Beef	
	Carcass[a]
	Ground
Pork	
	Carcass
Chicken	
	Carcass
	Parts
	Comminuted (including ground)
Turkey	
	Carcass
	Comminuted (including ground)
Products with *Campylobacter* standards	
Chicken	
	Carcass
	Parts
	Comminuted (including ground)
Turkey	
	Carcass
	Comminuted (including ground)

Source: GAO analysis of U.S. Department of Agriculture (USDA) documents. | GAO-18-272. Notes: Ground pork is a type of comminuted (i.e., broken into a number of pieces) product that also includes sausage and patties. Ground poultry is a type of comminuted product that also includes deboned products, among others. Chicken parts include breasts, legs, and wings. USDA originally selected Salmonella for pathogen standards because it was the most common cause of foodborne illness associated with meat and poultry products and because FSIS believed that intervention strategies aimed at reducing fecal contamination and other sources of Salmonella in raw products might be effective against other pathogens. According to USDA, *Campylobacter* is most often isolated from the intestinal tract of poultry and from poultry products.

[a]USDA has separate pathogen standards for cows/bulls and steers/heifers.

Figure 3. Raw Meat and Poultry Products with Pathogen Standards.

According to the December 2016 document, certain agency officials are to meet as needed to discuss emerging food safety risks and propose related data collection efforts to senior management, who will decide which products to consider for new standards. However, the document does not explain the basis for management's decisions. FSIS has informed stakeholders that it will take into account factors including consumption and foodborneillness data, as it did when setting standards for chicken parts, but the agency has not documented this process going forward. Several researchers and consumer advocacy representatives we interviewed questioned whether the agency's process proactively addresses food safety risks.

Previously, FSIS developed new pathogen standards after thc agency was directed to do so or after widespread outbreaks indicated the need. For example, in 2011, FSIS revised *Salmonella* standards for chicken and turkey carcasses and developed new standards for *Campylobacter* in these same products after being charged with doing so by the Presidential Food Safety Working Group.[28] Additionally, in a 2016 *Federal Register* notice,[29] FSIS, after reviewing outbreaks from these products in 2011, 2013, and 2015—outbreaks in which 794 people were sickened and 1 died—concluded that new pathogen standards were needed for comminuted (including ground and other mechanically separated) poultry and chicken parts.

Under federal standards for internal control, federal entities are to design control activities to achieve objectives and respond to risks, including appropriate documentation of transactions and internal control.[30] With appropriate documentation of internal control, management clearly documents internal control and allows the documentation to be readily available for examination; the documentation may appear in management directives, administrative policies, or operating manuals. Until FSIS clearly documents its process for deciding which products to consider for new

---

[28] In March 2009, the President established the Food Safety Working Group to coordinate federal efforts and establish food safety goals to make food safer. This working group served as a centralized mechanism for broad-based food safety collaboration and resulted in a number of accomplishments, including improved coordination. However, the working group is no longer meeting. See GAO, *Federal Food Safety Oversight: Additional Actions Needed to Improve Planning and Collaboration*, GAO-15-180 (Washington, D.C.: Dec. 18, 2014).

[29] 81 Fed. Reg. 7285 (February 11, 2016).

[30] GAO, *Standards for Internal Control in the Federal Government,* GAO-14-704G (Washington, D.C.: September 2014).

pathogen standards, including the basis on which such decisions should be made, FSIS will not have assurance that its decisions are risk-based and that agency personnel will know the process when making such decisions.

USDA Has Revised Some Existing Pathogen Standards But Has Not Revised Others in Decades and Has No Time Frames for Revision

USDA's FSIS has revised *Salmonella* standards for chicken and turkey carcasses and for comminuted chicken and turkey but has not revised other *Salmonella* standards since 1996, and the agency has not set time frames for determining whether revisions are needed. Specifically, as noted above, FSIS revised *Salmonella* standards for chicken and turkey carcasses in 2011 in response to a charge from the President's Food Safety Working Group that the agency develop new or revised standards to reduce the prevalence of *Salmonella* in poultry products. The agency revised the pathogen standards for comminuted chicken and turkey in 2016 to help achieve public health goals for reducing human illness from *Salmonella*, among other things.[31] The revisions have generally involved reductions in the maximum allowable percentage of products that test positive for this pathogen. For example, in 2016, when the agency revised the *Salmonella* standards for comminuted chicken, the allowable percentage changed from 44.6 to 25.0 (See Table 1).

However, FSIS has not revised the *Salmonella* standards for beef and pork carcasses and ground beef since they were first developed in 1996. Although USDA announced in a 2014 *Federal Register* notice that it intended to propose new pathogen standards for ground beef,[32] FSIS has not done so. Furthermore, FSIS set the pathogen standards for beef and pork carcasses and ground beef at industry-wide prevalence levels found at

---

[31] FSIS developed these goals, and associated targets, in coordination with other federal agencies, including CDC and FDA.

[32] 79 Fed. Reg. 32,436 (June 5, 2014).

that time,[33] not at levels intended to be protective of human health. In 2017, FSIS reviewed data on *Salmonella* in beef carcasses and ground beef and determined that the agency will not reach public health goals for reducing foodborne illness from *Salmonella* without further reduction in *Salmonella* contamination in beef. FSIS officials said that the agency is developing options for how it might move forward and could determine that revised or new standards are not needed and that other policies could suffice in addressing pathogens in beef. In the meantime, however, the agency in 2014 suspended monitoring against the existing *Salmonella* standards for ground beef until the agency develops a revised standard.[34] The agency also suspended monitoring whether plants were meeting the pathogen standard for *Salmonella* on pork carcasses because, according to agency officials, the percentage of pork carcass samples that tested positive for *Salmonella* was consistently low. FSIS officials said that the agency is collecting data on pathogens in pork that could lead to new standards for pork products.[35] In the absence of testing against the standards, the agency has other tools to ensure plants are controlling pathogens. For example, the agency continues to test beef for levels of *E. coli*, and FSIS inspectors at plants are to routinely check records to verify a plant's compliance with its HACCP plans. FSIS officials told us that they would begin monitoring against the *Salmonella* standards for these products if the standards are revised or determined to be sufficient (in the case of beef and pork carcasses and ground beef) or if the agency develops new standards (in the case of pork cuts and ground pork). Generally, FSIS begins monitoring against a standard once the agency announces a standard and after a phase-in period has ended.

---

[33] Prevalence is the proportion of a product that would test positive for a pathogen if the entire population of that product was sampled and analyzed during a specific period of time.

[34] Per agency officials, USDA was also to use those resources for other exploratory sampling programs that could result in revised standards for beef carcasses and ground beef or new standards for pork parts.

[35] In 2015, FSIS sampled various pork products for *Salmonella*, *Campylobacter*, *Toxoplasma gondii*, *Yersinia enterocolitica*, and other pathogens. In 2016, FSIS extended sampling for certain pork products.

Table 1. USDA Pathogen Standards and the Year Developed or Revised

Product	Maximum acceptable percentage of products sampled to test positive for *Salmonella*		Maximum acceptable percentage of products sampled to test positive for *Campylobacter*	
	Initial standard (year set)	Current standard (year revised)	Initial standard (year set)	Current standard (year revised)
Beef				
Steers/heifers carcass	1.0	1.0	n/a[d]	n/a[d]
	(1996)	(not revised)		
Cows/bulls carcass	2.7	2.7	n/a[d]	n/a[d]
	(1996)	(not revised)		
Ground	7.5	7.5a	n/a[d]	n/a[d]
	(1996)	(not revised)		
Pork				
Carcass	8.7	8.7	n/a[d]	n/a[d]
	(1996)	(not revised)		
Chicken				
Carcass	20	9.8	15.7	15.7
	(1996)	(2011)	(2011)	
Parts[b]	15.4	15.4	7.7	7.7
	(2016)		(2016)	
Comminuted[c]	44.6	25.0	1.9	1.9
	(1996)	(2016)	(2016)	
Turkey				
Carcass	19.6	7.1	5.4	5.4
	(2005)	(2011)	(2011)	
Comminuted[c]	49.9	13.5	1.9	1.9
	(1996)	(2016)	(2016)	

Legend: n/a = no pathogen standard developed. Source: GAO analysis of U.S. Department of Agriculture (USDA) documents. | GAO-18-272. Note: Pathogen standards set in 1996 for beef, pork, chicken carcasses, and ground beef are expressed as a prevalence level, i.e., the proportion of a product that would test positive for a pathogen if the entire population of that product was sampled and analyzed during a specific period. Pathogen standards set or revised in 2011 or 2016 are calculated as the percentage of samples with detectable levels of pathogens from a specified set of samples, which varies by pathogen standard.

[a]In a 2014 Federal Register notice, USDA announced that it intended to propose new Salmonella standards for ground beef. 79 Fed. Reg. 32,436 (June 5, 2014). USDA has not yet proposed such standards.

[b]Chicken parts includes breasts, legs, and wings.

[c]Comminuted chicken and turkey include ground and deboned products.

[d]A 2017 report from the Interagency Food Safety Analytics Collaboration—a tri-agency group created by the Centers for Disease Control and Prevention, the U.S. Food and Drug Administration, and USDA—on the percentage of foodborne illnesses related to Campylobacter in 2013 attributed 29.2 percent of foodborne illnesses to chicken, 6.5 percent to pork, 4.9 percent to turkey, 1.6 percent to beef, and 12 other food groups accounted for the remainder.

For example, when FSIS developed new *Campylobacter* and *Salmonella* standards for chicken parts in 2016, the agency began monitoring whether plants met the standard 5 months after the standards were announced in the *Federal Register*. Monitoring for compliance with pathogen standards is a key tool as envisioned by the 1996 Pathogen Reduction; HACCP Systems final rule for verifying the effectiveness of a plant's processing controls to prevent, eliminate, or reduce food safety hazards. It is unclear when FSIS plans to resume the use of this tool and complete the revisions of the *Salmonella* standards for beef carcasses or ground beef or develop new standards for additional pork products because the agency has not set time frames for doing so. According to FSIS officials, developing or revising pathogen standards takes time and resources, in part because the agency must first collect and analyze data to estimate the prevalence of pathogens in FSIS-regulated products, notify the public of proposed standards, and open a comment period, all of which can take years. However, according to FSIS officials, the agency has no time frames for determining what actions to take. Program schedule planning is recognized as a leading practice to ensure organizational activities are completed as planned, according to the Project Management Institute's Standard for Program Management. Such planning includes setting time frames for completing a project. By setting time frames for determining what pathogen standards or additional policies are needed to address pathogen levels in beef carcasses, ground beef, and pork products, FSIS could better ensure it completes these activities in a timely manner to protect human health.

USDA Is Taking Additional Steps to Address Pathogen Reduction Challenges That We Identified in 2014, But These Challenges Are Ongoing

In addition to taking steps to develop or revise pathogen standards, USDA's FSIS is addressing other challenges we identified in September

2014 with respect to poultry pathogens, but these challenges are ongoing and also apply to meat products. These challenges include FSIS's limited control over factors that affect the level of pathogens outside of plants, pathogens not designated as hazards, the complex nature of *Salmonella*, limited *Campylobacter* research and testing, limited enforcement authority,[36] absence of mandatory recall authority, and insufficient prevalence estimates.

Limited Control Outside of Plants

Limited Control Outside of Plants
In September 2014, we found that the U.S. Department of Agriculture's (USDA) Food Safety and Inspection Service (FSIS) faced a challenge in reducing levels of Salmonella and Campylobacter in poultry products in part because the agency did not have regulatory jurisdiction over farm practices to reduce contamination in poultry before they reach a plant for slaughter and processing. At the time, we noted that FSIS had worked to address the on-farm limitation by issuing guidelines that detailed, among other things, several on-farm practices to reduce Salmonella and Campylobacter in live poultry. We recommended that in future revisions of the guidelines, FSIS include information on the effectiveness of on-farm practices to explain the potential benefits of adopting such practices on poultry farms. USDA concurred with our recommendation. In addition, we found that once poultry products leave a plant, factors beyond FSIS's control may affect contamination of poultry products, such as cross-contamination from poultry products (i.e., when bacteria spread from a food to a surface, from a surface to another food, or from one food to another) that can occur at retail establishments, in restaurants, and in consumers' homes, according to a food safety researcher we interviewed.

Source: GAO | GAO-18-272.

Since our September 2014 report, FSIS has taken steps to help overcome not having regulatory jurisdiction to reduce the level of pathogens before and after slaughter and processing. With respect to reducing pathogens before slaughter and processing, the agency has implemented our recommendation to include information on effectiveness in its guidelines about on-farm practices to reduce *Salmonella* and *Campylobacter* in live poultry. The agency has also included such

[36] We identified limited enforcement authority as a potential challenge, rather than a challenge, because FSIS officials told us that the agency had tools to overcome enforcement authority limitations.

information in guidelines for beef cattle.[37] However, it has not taken similar action for hogs. Specifically:

- Poultry: FSIS drafted revised guidelines in 2015 regarding, among other things, on-farm practices for reducing *Salmonella* and *Campylobacter* in live chickens and turkeys, with information on the effectiveness of these practices, as a USDA advisory committee recommended in 2011 (for livestock and poultry) and as we recommended in September 2014 for poultry.[38] According to FSIS officials, the agency has kept the poultry guidelines in draft form while it responds to public comments on the document.
- Beef: The agency has already taken similar action to reduce *Salmonella* in beef cattle. Specifically, in 2014, FSIS revised guidelines regarding on-farm practices for reducing Shiga-toxin-producing *E. coli* in cattle—practices that, according to agency officials, may also assist in reducing *Salmonella*. These guidelines included information about the application of on-farm practices and research on their effectiveness. For example, the 2014 guidelines describe research on the effects of different feed—such as hay, grain, and grass—on reducing pathogens in cattle. In 2017, FSIS issued revised guidelines that also include information from the 2014 document on the effectiveness of on-farm practices.
- Pork: FSIS has had draft guidelines in place since 2013 concerning, among other things, on-farm practices to reduce levels of *Salmonella* contamination in hogs. The draft *Salmonella* guidelines are available on the agency's website. Even though the guidelines are not yet finalized, FSIS encourages producers to use them, according to agency officials we interviewed. However, unlike the poultry and beef cattle guidelines, the draft *Salmonella* guidelines do not contain information on the effectiveness of on-

[37] According to USDA data on livestock slaughtered under federal inspection, less than 10 percent of beef is from dairy cows. The agency's guidelines for beef cattle, which includes information regarding on-farm practices and the effectiveness of these practices, also includes some information for dairy cattle.

[38] GAO-14-744.

farm practices, as recommended in 2011 by USDA's National Advisory Committee on Meat and Poultry Inspection.[39] According to the draft guidelines, when a plant makes changes at the appropriate processing location, process control should result in raw pork products that have less contamination with pathogens, including *Salmonella*. FSIS officials we interviewed told us that there is not as much research available for such practices for hogs as there is for beef cattle and poultry. However, the officials agreed that including available information would be beneficial. By including available information on the effectiveness of these practices to reduce the level of pathogens as it finalizes its guidelines for controlling *Salmonella* in hogs, FSIS would have better assurance that it is keeping industry informed of the potential benefits of adopting on-farm practices and encourage their implementation.

With respect to reducing pathogens after slaughter, FSIS continues to update its guidance to consumers and work with federal partners to ensure the safety of meat and poultry products after they leave the plant. For example:

- In 2015, the agency developed the FoodKeeper mobile application to educate consumers on how to use food while at peak quality and store food properly.[40] It updated the application in 2017 so users could receive automatic notifications when FDA or FSIS announces food safety recalls.

---

[39] Established in 1971, the National Advisory Committee on Meat and Poultry Inspection advises the Secretary of Agriculture on matters affecting federal and state inspection program activities. National Advisory Committee on Meat and Poultry Inspection, *Final Report – NACMPI Recommendations for FSIS on Pre-Harvest Salmonella* (Washington, D.C.: September 2011).

[40] The application provides consumers with information on proper cooking temperatures and how long food can be stored.

Pathogen Contamination after Products Leave the Plant

According to the Centers for Disease Control and Prevention (CDC), even if meat and poultry products leave the processing or slaughter plant with no detectable pathogen, it does not ensure that the products are safe, as opportunities exist for them to become contaminated at any point along the farm-totable continuum. To illustrate, frozen hamburger patties might be trucked from a plant to a supplier, stored in the supplier's warehouse for a few days, trucked again to a local distribution facility, and then delivered to a restaurant. According to CDC, if refrigerated food is left on a loading dock during transportation for an extended time in warm weather, the food could reach temperatures that allow pathogens to grow.

Contamination can also occur during preparation in consumers' homes if food is not properly stored, prepared, heated, or served. For example, according to CDC, once contamination occurs, if meat and poultry are stored or cooked at unsafe temperatures, pathogens will grow quickly, which may lead to foodborne illness.

Source: GAO analysis of information provided by CDC. Top image: Courtesy of U.S. Department of Agriculture (USDA), Lance Cheung. Bottom image: Courtesy of USDA. | GAO-18-272.

- In 2016, FSIS and FDA announced that they would work together to revise the FDA Food Code—a model that local, state, tribal, and federal regulators use to ensure food safety at retail stores, restaurants, and institutions such as nursing homes, among others— to ensure consistency with FSIS regulations and guidance.[41]

---

[41] According to its website, FDA published the Food Code in its current format every 2 years from 1993 to 2001. With the support of the Conference for Food Protection, FDA decided to move to a 4-year interval between complete Food Code editions. During the interim period between full editions, FDA may publish a Food Code Supplement that updates,

- In 2017, FSIS expanded the operating hours for its Meat and Poultry Hotline, through which consumers could speak with an agency representative or listen to recorded messages regarding food safety, such as the proper storage, handling, and preparation of meat and poultry products.

Pathogens Not Designated as Hazards

Pathogens Not Designated as Hazards

In September 2014, we found that the U.S. Department of Agriculture's Food Safety and Inspection Service (FSIS) faced a challenge in reducing Salmonella and Campylobacter contamination in poultry products when plants do not designate these pathogens as hazards. Under the Hazard Analysis and Critical Control Point (HACCP) approach, plants have discretion about whether to include Salmonella or Campylobacter as a hazard "reasonably likely to occur" in their HACCP plans and develop mitigation strategies to reduce these pathogens. FSIS's 2014 final rule for modernizing poultry slaughter inspection requires plants to develop, implement, and maintain written procedures to prevent contamination of carcasses and parts by enteric pathogens—bacteria that normally reside in the intestines of many animals, including humans, such as Salmonella and Campylobacter—as well as fecal material.[42] Plants must incorporate these procedures into their HACCP plans, sanitation procedures, and other programs.

Source: GAO | GAO-18-272.

Since our September 2014 report, FSIS has not required hog and beef plants to designate *Salmonella* or *Campylobacter* as hazards likely to occur, but it has taken other steps to reduce *Salmonella* and *Campylobacter* contamination when plants do not designate these pathogens as hazards.[43] More specifically, in February 2018, FSIS proposed a rule to modernize hog slaughter inspections. The proposed rule would require plants to

---

modifies, or clarifies certain provisions. The 2005 Food Code was the first full edition published on the new 4-year interval, and it was followed by the Supplement to the 2005 Food Code, which was published in 2007. The 2013 Food Code is the most recent full edition published by FDA.

[42] 79 Fed. Reg. 49,566 (Aug. 21, 2014).

[43] A 2017 report from the Interagency Food Safety Analytics Collaboration, a tri-agency group created by CDC, FDA, and FSIS, on the percentage of foodborne illness related to *Campylobacter* in 2013 attributed 29.2 percent of foodborne illnesses to chicken, 6.5 percent to pork, 4.9 percent to turkey, 1.6 percent to beef, and 12 other food groups accounted for the remainder.

develop, implement, and maintain written procedures to prevent contamination by enteric pathogens in pork.

Stakeholders we interviewed representing industry and consumer advocacy groups disagreed on whether plants should be required to designate specific pathogens as a hazard reasonably likely to occur. However, in response to instances in which inadequate validation of HACCP plans led to the production of adulterated food, and in some cases illnesses, FSIS released compliance guidance outlining best practices for designing and implementing adequatc IIACCP plans for all plants in 2015. According to FSIS, plants can use the guidance to properly design and execute HACCP plans and reduce the likelihood of contamination of the products they produce. Specifically, the guidance outlines, among other things, best practices for gathering scientific and technical support, as part of the HACCP plan validation process, to demonstrate that the plants' processes prevent, reduce, or eliminate the hazards identified.

Complex Nature of *Salmonella*

Complex Nature of *Salmonella*

In September 2014, we found that the U.S. Department of Agriculture's Food Safety and Inspection Service (FSIS) faced a challenge in reducing *Salmonella* contamination in poultry products because of the complex nature of this pathogen. The majority of the representatives from industry and consumer groups we interviewed at the time, as well as FSIS officials, agreed that *Salmonella* is difficult to control in poultry products because it is widespread in the natural environment. According to Centers for Disease Control and Prevention officials we interviewed for our past work, there are more than 2,500 serotypes of *Salmonella* (with different strains), some of which pose greater risk to human health than others. Therefore, it is important to understand the genetic makeup of each to determine which ones are more or less likely to cause human illness.

Source: GAO | GAO-18-272.

FSIS officials said that, in the future, there may be opportunities to improve how the agency protects human health by focusing inspections on plants and products that have tested positive for the more dangerous strains of *Salmonella* in meat and poultry products.[44] To this end, FSIS collaborates with USDA's Agricultural Research Service and APHIS, CDC, FDA, and local and state public health partners to develop new technologies that can more precisely determine if a strain of *Salmonella* detected is particularly dangerous to people. One such technology is whole genome sequencing, which allows the agency to determine the complete set of genes, or strain, within a *Salmonella* serotype.[45]

According to FSIS officials, it is more challenging to link the strain associated with an illness to a specific meat or poultry product that has sickened consumers; whole genome sequencing technology can more definitively identify the strain involved in an outbreak and help reduce incidents of illness or death due to foodborne pathogens. FSIS is currently planning how to integrate this technology into its food safety program. For example, current pathogen standards are based on the presence or absence of generic *Salmonella*, not on specific strains. FSIS held a public meeting in October 2017 to get input from state, federal, and international public health partners and other stakeholders on the use of this technology in a regulatory setting to improve food safety and public health.

---

[44] A strain refers to a specific subtype of a microorganism, such as a bacterium or virus, that has distinctive traits in its shape or genes that separate it from other strains of the same microorganism. A serotype is one way to group strains that share distinctive surface structures. For example, *Salmonella* bacteria look alike under the microscope but can be separated into many serotypes based on two structures on their surface. These surface structures reflect the antigenic composition—molecular structures that are recognized by the immune system and are capable of triggering an immune response—of a strain.

[45] Whole genome sequencing (also known as full genome sequencing, complete genome sequencing, or entire genome sequencing) is the process of determining the complete DNA sequence of an organism's genome at a single time. This technique allows detailed comparisons among samples. According to FSIS officials, the agency typically uses sequencing technology to identify a sequence, but not the whole genome. However, this is still referred to as "whole genome sequencing." According to CDC, whole genome sequencing is like comparing all of the words in a book, instead of just the number of chapters, to see if the books are the same or different. In March 2017, we reported on efforts by USDA, CDC, and FDA to use whole genome sequencing to identify and report on antibiotic-resistant bacteria through the National Antimicrobial Resistance Monitoring System. See GAO, *Antibiotic Resistance: More Information Needed to Oversee Use of Medically Important Drugs in Food Animals,* GAO-17-192 (Washington, D.C.: Mar. 2, 2017).

Limited *Campylobacter* Research and Testing

Limited *Campylobacter* Research and Testing
In September 2014, we found that the U.S. Department of Agriculture's Food Safety and Inspection Service faced a challenge in reducing levels of *Campylobacter* in poultry products in part because less was known about *Campylobacter* than about *Salmonella*. Specifically, technologies, such as clinical diagnostics, used to detect *Campylobacter* may have underdiagnosed cases of illness from this pathogen, and the methods used by many diagnostic laboratories to isolate *Campylobacter* from samples were not standardized, according to a 2012 World Health Organization report on illnesses from the pathogen. Additionally, the agency's ability to measure a reduction in *Campylobacter* illnesses depended on its ability to attribute *Campylobacter* illnesses to poultry and other food types, according to agency officials, and attribution analyses needed improvement.

Source: GAO | GAO-18-272.

Since our report in September 2014,[46] FSIS has had efforts under way with other agencies to improve foodborne illness source attribution to meat and poultry products and has independent data collection efforts under way to determine the presence of *Campylobacter* on these products. More specifically, in collaboration with CDC and FDA through the Interagency Food Safety Analytics Collaboration, FSIS has taken steps to improve and standardize methods to estimate the source attribution for *Campylobacter* foodborne illness. In 2015, this interagency collaboration improved the method for estimating the number of *Campylobacter* illnesses from meat and poultry products by standardizing the approach used by all three food safety agencies. The interagency collaboration's new estimates for the proportion of *Campylobacter* illnesses included all food products—including beef, pork, and poultry. The interagency collaboration also released updated foodborne illness source attribution estimates in December 2017. According to FSIS officials, the three agencies are collaborating on multiple analytic projects, in line with the interagency collaboration's 2017–2021 strategic plan, to improve models to estimate foodborne illnesses from *Campylobacter* and other pathogens. These projects involve using new methods and whole genome sequencing and other data sources.

---

[46] GAO-14-744.

In addition to this interagency effort, in 2015, FSIS tested about 200 samples of pork products for *Campylobacter* as part of an exploratory sampling effort, according to agency documents summarizing the efforts. FSIS found that about 1 percent of products tested were positive for *Campylobacter* and, therefore, chose not to continue testing pork products for this pathogen.[47] For poultry, in 2016, FSIS revised a laboratory guidebook describing standard protocols for isolating and analyzing *Campylobacter* in raw products. In 2017, the agency concluded a literature review of *Campylobacter* contamination in beef and, as of October 2017, is discussing the development of an exploratory sampling project to test for *Campylobacter* in beef products, according to agency officials.

Limited Enforcement Authority

Since our report in September 2014, FSIS continues to stand by the position that the 2000 court ruling does not affect its ability to use pathogen standards as a tool to prevent contaminated products from entering the market. FSIS reaffirmed its position in a 2016 *Federal Register* notice.[48] Our review of FSIS data from 2016 through 2017 for poultry plants shows that some plants are still not meeting pathogen standards—in some cases repeatedly not meeting the standards—and are allowed to operate.[49] We were unable to review similar data for beef or hog plants since, as noted above, FSIS suspended monitoring these plants against pathogen standards. FSIS stands by its assessment that its

---

[47] We did not examine whether FSIS's sample was sufficient to determine the risk of *Campylobacter* in pork products.

[48] 81 Fed. Reg. 7,285 (Feb. 11, 2016).

[49] Because FSIS does not test against the beef and pork standards, we reviewed FSIS data to determine the extent to which plants were not meeting chicken carcasses pathogen standards. Specifically, our review of FSIS monthly data from May 2016 through August 2017 on whether plants producing chicken carcasses met the pathogen standards showed that 50 plants of 210 did not meet the *Salmonella* pathogen standard for at least 1 month of the data reporting period, and 20 plants of 210 did not meet the *Campylobacter* standard for at least 1 month in the same period. In some cases, these plants did not meet the standard for the entire period that we reviewed: 2 plants did not meet the *Salmonella* standard and 4 plants did not meet the *Campylobacter* standard at all during this period.

enforcement tools are sufficient. Moreover, in 2015, FSIS announced an additional tool to help FSIS identify and assess problems or trends that may be of concern.

Limited Enforcement Authority

In September 2014, we found that the U.S. Department of Agriculture's Food Safety and Inspection Service (FSIS) faced a potential challenge in reducing Salmonella contamination in poultry products, according to agency officials and representatives of some stakeholder groups we interviewed, because (1) a 2000 federal court ruling stated that FSIS could not withdraw inspectors, effectively shutting down the plant, solely because a plant did not meet Salmonella pathogen standards,[50] and (2) FSIS has not classified Salmonella as an adulterant in raw poultry products, so products contaminated with this pathogen generally may be permitted to enter commerce.

FSIS adopted the position that the court ruling did not affect its ability to use the Salmonella pathogen standards as part of verifying a plant's sanitation and Hazard Analysis and Critical Control Point plans and that it had tools, such as food safety assessments (an evaluation of a plant's food safety system), to prevent contaminated products from entering the market. Representatives from consumer groups we interviewed at the time said that even with these tools, the agency does not have sufficient authority to ensure plants comply with the standards because FSIS cannot shut down plants when they fail the Salmonella standards alone. Representatives from industry groups we interviewed at the time disagreed and stated that FSIS has sufficient authority to ensure plants comply with standards because the agency has broad statutory authority and oversight.

Regarding FSIS not classifying Salmonella as an adulterant, representatives from consumer groups we interviewed for our previous work said that the agency should declare some serotypes of Salmonella as adulterants, such as those with specific antibiotic-resistant patterns. FSIS officials we interviewed for our previous work said they found no conclusive evidence that antibiotic-resistant strains of Salmonella or Campylobacter have a greater resistance to the interventions used in plants but that the agency would continue to review relevant scientific evidence to identify any potential challenges these serotypes may present to public health.

Source: GAO | GAO-18-272.

Specifically, FSIS investigators must now conduct a public health risk evaluation at every plant that does not meet a pathogen standard. This is a positive step for those products that have pathogen standards, such as

[50] Supreme Beef Processors v. U.S. Dep't of Agric., 113 F. Supp. 2d 1048 (N.D. Tex. 2000). A federal appeals court upheld this decision in 2001. The appeals court held that USDA's *Salmonella* standards conflict with the agency's authority to enforce the prohibition against meat or meat products prepared, packed, or held in insanitary conditions and the agency could not use *Salmonella* verification tests results on products alone to determine whether a plant was in compliance with sanitary requirements. Supreme Beef Processors v. U.S. Dep't of Agric. 275 F. 3d 432 (5th Cir. 2001). On a day-to-day basis, if an FSIS inspector finds plant conditions or procedures insanitary, the inspector can refuse to perform inspection and temporarily suspend the plant's operation until the problem is corrected.

chicken parts. However, as previously stated, FSIS does not test for whether plants producing beef carcasses, ground beef, and pork carcasses meet the pathogen standards for those products, and other products such as ground pork do not have pathogen standards. Representatives from consumer groups and industry we interviewed continue to disagree on whether FSIS' existing enforcement tools are sufficient to ensure that meat and poultry plants meet pathogen standards.[51]

Regarding antibiotic-resistant strains of *Salmonella*, FSIS officials continue to state that the pathogen does not meet the criteria for classifying it as an adulterant and that the agency will continue to examine options for regulating the presence of antibiotic-resistant strains of *Salmonella* in raw meat and poultry products.[52] Agency officials told us that to classify a pathogen as an adulterant in raw meat and poultry products, FSIS must determine that the pathogen meets certain criteria established both in its authorizing statutes and by case law. Specifically, in *American Public Health Association v. Butz*, a federal appeals court in 1974 held that *Salmonella* did not adulterate raw poultry because ordinary consumer methods of preparing and cooking the product would eliminate the pathogen. In contrast, FSIS declared certain types of *E. coli* as adulterants in beef, as discussed above, because ordinary consumer cooking does not eliminate the pathogen. According to FSIS officials, the available data do not appear to indicate that *Salmonella* presents the same issues as *E. coli* or meets the necessary criteria, regardless of whether it is resistant or susceptible to antibiotics. This issue continues to be contentious among the stakeholders we interviewed. Six of the seven industry stakeholders we

---

[51] Specifically, 8 of the 17 stakeholders we interviewed (6 stakeholders from industry and 2 food safety researchers) told us that FSIS's enforcement authority is sufficient. In contrast, 9 of the 17 stakeholders we spoke with (1 stakeholder from industry, 4 stakeholders from consumer advocacy groups, 2 food safety researchers, and 2 former federal food safety experts) stated that FSIS's enforcement tools are limited because the agency cannot legally shut down a plant for poor performance, among other reasons.

[52] For information about USDA's efforts to address antibiotic resistance in livestock and poultry, see GAO, *Antibiotic Resistance: Federal Agencies Need to Better Focus Efforts to Address Risk to Humans from Antibiotic Use in Animals,* GAO-04-490 (Washington, D.C.: Apr. 22, 2004); *Antibiotic Resistance: Agencies Have Made Limited Progress Addressing Antibiotic Use in Animals,* GAO-11-801 (Washington, D.C.: Sept. 7, 2011); and *Antibiotic Resistance: More Information Needed to Oversee Use of Medically Important Drugs in Food Animals,* GAO-17-192 (Washington, D.C.: Mar. 2, 2017).

interviewed stated that FSIS's current enforcement authority is sufficient. Two of four food safety researchers we interviewed stated that the agency does not need additional authority to label *Salmonella* as an adulterant because FSIS has labeled other pathogens as adulterants when it made sense to do so, such as *E. coli*¸ and there is no need to label naturally occurring bacteria as adulterants on raw product. In contrast, all four of the consumer advocacy groups and two of the four food safety researchers we interviewed stated that FSIS needs more authority to label *Salmonella* as an adulterant.

No Mandatory Recall Authority

No Mandatory Recall Authority
In September 2014, we found that the U.S. Department of Agriculture's Food Safety and Inspection Service (FSIS) faced a challenge in reducing Salmonella and Campylobacter contamination in poultry products because it did not have mandatory food recall authority similar to that of the Food and Drug Administration (FDA) for the food products FDA regulates, such as milk, seafood, fruits, and vegetables. In 2011, Congress passed the FDA Food Safety Modernization Act, giving FDA mandatory recall authority. We recommended in October 2004 that Congress consider legislation to increase FSIS's authority to include mandatory recalls, but the agency continues to not have such authority.[53] Instead, to protect human health from potentially contaminated meat and poultry products, FSIS can issue public health alerts, which notify the public on specific actions to take to avoid illness, or request voluntary recalls, which are voluntary actions taken by plants, among other actions. Before requesting a voluntary recall, FSIS must gather sufficient evidence through its investigation and determine that a product is adulterated and mislabeled, among other things. In September 2014, we reported that this can be challenging to do. FSIS officials told us at the time that rather than focusing on the lack of mandatory recall authority, it was more productive to work aggressively with the tools they had, such as withdrawing inspectors, thus preventing products from entering commerce. According to FSIS officials, this can be as effective for keeping unsafe food from the marketplace as FDA's recall authority.

Source: GAO | GAO-18-272.

---

[53] GAO, *Food Safety: USDA and FDA Need to Better Ensure Prompt and Complete Recalls of Potentially Unsafe Food,* GAO-05-51 (Washington, D.C.: Oct. 6, 2004).

Since our September 2014 report, FSIS officials said that they continue to believe that mandatory recall authority is not necessary for the reasons previously mentioned.

According to FSIS officials, the agency continues to refine and improve its procedures for requesting voluntary recalls of adulterated and misbranded meat and poultry products, confirming the effectiveness of these recalls, and alerting the public about adulterated and misbranded products that may remain in commerce.[54] Therefore, FSIS officials stated that the agency does not see the lack of mandatory recall authority as an obstacle or hindrance to its efforts to protect public health and ensure that meat and poultry products are safe, wholesome, and properly labeled. In contrast, FDA officials told us that having mandatory recall authority protects human health from foodborne illness because the agency does not have to rely upon manufacturers' voluntary recall efforts or obtain a court order to remove contaminated or misbranded food, other than infant formula, from the food supply. In our review of FDA's annual reports to Congress on the use of mandatory recall authority from 2013 to 2016, the most recent available, the agency has used its mandatory recall authority twice. The majority (12 of 17) of the stakeholders we interviewed stated that the absence of mandatory recall authority is not a challenge for FSIS in reducing pathogen contamination of meat and poultry products. However, according to 3 of 4 stakeholders from consumer groups and 1 of 4 food safety researchers we interviewed, acquiring mandatory recall authority would enable FSIS to better protect human health because the agency would then have an additional tool to stop an outbreak of foodborne illness and address the level of pathogens in products once they leave the plant.

---

[54] Voluntary recalls are voluntary actions taken by plants under the direction of their company. When companies discover that they may have distributed food that is contaminated with disease-causing bacteria or that contains allergens that can cause serious illness or death, they may conduct a voluntary recall. That is, they will contact their customers and instruct them to contact the wholesalers, retailers, and others in the food's distribution chain and ask them to return or destroy the potentially unsafe food. If a plant conducts a recall, FSIS provides assistance and monitors the recall. If a plant does not conduct a recall that FSIS has requested, FSIS is limited to its authority to detain and seize the products in question.

Insufficient Prevalence Estimates

In 2016, FSIS implemented this new testing approach for all poultry products for which there are pathogen standards and for some meat products, but according to officials, the agency did not do so for all products that it regulates because of resource constraints. Specifically, according to a 2016 *Federal Register* notice, FSIS now routinely samples chicken and turkey carcasses, chicken parts (legs, wings, and breasts), and comminuted chicken and turkey for *Salmonella* and *Campylobacter* pathogens over an entire year—rather than a set period of time—based on the volume of poultry products produced in plants.[55] It also uses this approach to test for *Salmonella* in ground beef, beef manufacturing trimmings, and other ground beef components, according to a 2014 *Federal Register* notice.[56] This new approach allows for better prevalence estimates and for monitoring changes in prevalence over time, according to agency officials. As discussed earlier, FSIS began exploratory sampling of pork products, including pork cuts and comminuted (including ground) pork, in 2015. According to a 2017 agency notice describing the sampling, FSIS collects and analyzes samples of pork products in a way that allows for prevalence estimates. FSIS does not use the same approach to sample other products, such as raw components used in ground beef (e.g., esophagus, head meat, cheek meat, and hearts), chicken half carcasses, and chicken necks, because of limited resources, according to agency officials. These officials stated that the agency first conducts exploratory sampling—such as its current program for pork products—to determine if FSIS should allocate resources for routine sampling of these products that would allow for prevalence estimates.

[55] 81 Fed. Reg. 7,285 (Feb. 11, 2016).
[56] 79 Fed. Reg. 32,436 (Jun. 5, 2014).

USDA's Release of Plant Performance Information

To encourage poultry slaughter and processing plants to control for Salmonella and Campylobacter—disease-causing pathogens that can sicken consumers—USDA publicly releases information on individual plant performance for reducing these pathogens. According to the agency's 2017 annual plan, publishing plant-specific data allows consumers to make more informed choices, motivates individual plants to improve performance, and leads to industry-wide improvements in food safety. USDA's Economic Research Service found that publicly releasing the identities of plants with poor or mediocre performance on tests for Salmonella is strongly correlated with about a 60 percent decline of chicken carcass samples testing positive for Salmonella from 2006 to 2010. In 2016, USDA temporarily replaced posting information on individual plants' performance for chicken and turkey carcasses, chicken parts, and comminuted poultry (e.g., ground), with information on aggregate results to allow time for plants to update their food safety systems. In January 2018, FSIS began reposting individual plants' category status for poultry carcasses on a monthly basis. According to the agency's annual plan for fiscal year 2017, USDA intends to resume publicly releasing individual plant performance information for turkey carcasses and to add data for plants producing chicken parts and comminuted chicken and turkey. The agency also intends to release data for plants producing some beef products, according to its 2016 strategic plan on publicly releasing data.

Source: GAO analysis of information provided by the U.S. Department of Agriculture (USDA). Image: Courtesy of USDA, Alice Welch. | GAO-18-272.

Insufficient Prevalence Estimates

In September 2014, we found that the U.S. Department of Agriculture's (USDA) Food Safety and Inspection Service (FSIS) faced a challenge in reducing Salmonella and Campylobacter contamination in poultry products as a result of not having sufficient prevalence estimates. Prevalence is the proportion of a product that would test positive for a pathogen if the entire population of that product was sampled and analyzed during a specific period of time. FSIS collects and analyzes data to estimate the prevalence of pathogens when the agency develops or revises pathogen standards for products it regulates. However, we reported that there were numerous problems with the data FSIS used to estimate prevalence. For example, assessing levels of poultry pathogens across the entire industry was difficult using data from FSIS's verification-testing program because the program was not designed to assess prevalence of pathogens industry-wide and the agency does not randomly select plants for inspection. According to USDA's National Advisory Committee on Microbiological Criteria for Food, estimating the prevalence of pathogens in food is critical to understanding and addressing the public health risk of foodborne illness, and these estimates provide a mechanism for measuring performance against public health goals, among other things. FSIS officials told us at that time that the agency had plans to propose a new testing approach for all of its poultry products, which would allow for more frequent data collection and improve prevalence estimates, among other things.

Source: GAO | GAO-18-272.

Conclusion

To help ensure the safety of meat and poultry products and protect against foodborne illness, USDA's FSIS has transitioned to an increasingly science-based, data-driven, risk-based approach. As part of this approach, FSIS has taken several actions to reduce levels of *Salmonella* and *Campylobacter* in poultry products, including strengthening existing pathogen standards for *Salmonella* in poultry carcasses and developing new *Salmonella* and *Campylobacter* standards for certain chicken parts. However, the agency has not set pathogen standards for many widely available products, such as pork cuts and ground pork, and the agency's process for deciding which products to consider for new pathogen standards is not fully documented. Previously, FSIS has developed new pathogen standards after the agency has been directed to do so or after widespread outbreaks indicated the need. Until FSIS clearly documents its process for deciding which products to consider for new pathogen standards, including the basis on which such decisions should be made, FSIS will not have assurance that its decisions will be risk-based and that agency personnel will know the process when making such decisions.

As part of its new approach, FSIS is collecting data that could enable it to set new pathogen standards for pork cuts and ground pork, and the agency is analyzing data that could lead to revising the *Salmonella* standards for beef carcasses and ground beef—which are decades old and not set at levels that are health protective. However, the agency has not set time frames for completing these efforts. In the absence of pathogen standards against which the agency tests, the agency is not using a valuable tool that could be used to help verify that plants' processing controls to prevent, eliminate, or reduce food safety hazards are working. By setting time frames for determining what pathogen standards or additional policies are needed to address pathogens in these products, FSIS could better ensure it completes these activities in a timely manner to better protect human health.

In addition, FSIS continues to face several challenges that hinder its ability to reduce the level of pathogens in meat and poultry products. For example, practices outside the slaughter plant, such as conditions on cattle, hog, and poultry farms, can affect levels of pathogens on meat and poultry products. To help overcome this challenge, the agency has developed draft guidance on practices for controlling levels of *Salmonella* and *Campylobacter* on beef cattle, hog, and poultry farms, but the draft guidance for hogs does not include available information on the effectiveness for each practice, as an internal agency committee recommended. As FSIS finalizes this guidance, FSIS could better inform industry of the potential benefits of adopting on-farm practices and encourage implementation of these practices by including available information on their effectiveness.

RECOMMENDATIONS FOR EXECUTIVE ACTION

We are making three recommendations to FSIS. Specifically:

The Administrator of FSIS should document the agency's process for deciding which products to consider for new pathogen standards, including the basis on which such decisions should be made. (Recommendation 1).

The Administrator of FSIS should set time frames for determining what pathogen standards or additional policies are needed to address pathogens in beef carcasses, ground beef, pork cuts, and ground pork. (Recommendation 2).

The Administrator of FSIS should include available information on the effectiveness of on-farm practices to reduce the level of pathogens as it finalizes its guidelines for controlling *Salmonella* in hogs. (Recommendation 3).

AGENCY COMMENTS AND OUR EVALUATION

We provided a draft of this chapter to USDA and the Department of Health and Human Services. In written comments, reproduced in appendix

II, USDA agreed with our three recommendations and described actions it will take to implement them.

In particular, with respect to our first recommendation, USDA stated that FSIS will complete an internal document that delineates the agency's process for creating or updating pathogen standards. However, USDA stated that although it agrees it can take additional steps to document its process, it does not agree that FSIS does not have assurance its decisions are risk based. In particular, it cited a *Federal Register* notice indicating that it designed its pathogen standards for chicken parts and comminuted chicken and turkey to achieve certain reductions in illnesses from *Salmonella* and *Campylobacter*. USDA also stated that FSIS has consistently documented and published its process in the *Federal Register*, and it noted that agency personnel use these *Federal Register* notices as guidance and historical reference. While these notices can be a useful historical record and document the steps FSIS took to ensure that agency decisions were risk-based, we continue to believe that, until FSIS clearly documents its process for deciding which products to consider for new pathogen standards going forward—including the basis on which such decisions should be made—FSIS will not have assurance that its decisions will be risk-based and that agency personnel know the process when making such decisions in the future. Completing documentation of the agency's process would address our recommendation.

Concerning our second recommendation, USDA stated that in 2018 FSIS will continue to assess data from sampling projects, along with baseline data and outbreak/illness data, to determine whether new or revised standards or additional policies are needed to address *Salmonella* in beef products. USDA further stated that in 2019, it will use data collected during its raw pork exploratory study to determine whether standards or additional policies (e.g., training, guidance to industry, or instructions to field personnel) are needed to address *Salmonella* in pork products. Finalizing analysis of these data and determining if additional standards or policies are needed to address *Salmonella* in beef in 2018 or pork in 2019 would address our recommendation.

In response to our third recommendation, USDA stated that FSIS will include available scientific information on the effectiveness of each recommended farm practice in the guidelines for reducing *Salmonella* in market hogs. Doing so would address our recommendation.

USDA also provided technical comments. We incorporated these comments as appropriate. The Department of Health and Human Services did not have any comments.

As agreed with your offices, unless you publicly announce the contents earlier, we plan no further distribution of this chapter until 30 days from the report date. At that time, we will send copies of this chapter to the appropriate congressional committees, the Secretary of Agriculture, the Administrator of the Food Safety and Inspection Service, and other interested parties.

Steve D. Morris
Director, Natural Resources and Environment.

APPENDIX I: *SALMONELLA* TESTING FOR BEEF PURCHASED FOR THE NATIONAL SCHOOL LUNCH PROGRAM

In addition to regulating meat and poultry sold in commerce, the U.S. Department of Agriculture (USDA) also purchases food and, in some cases, has additional food safety requirements for food it purchases.

USDA's Agricultural Marketing Service (AMS) purchases beef and other food for various federal nutrition assistance programs, including the National School Lunch Program.[57] USDA provides this food to states in support of about 100,000 public and private nonprofit schools that provide lunches to about 30 million children. Ground beef is a staple of school

[57] AMS purchases for the National School Lunch Program under authority of the Richard B. Russell National School Lunch Act. Pub .L. No. 79 P.L. 396, 60 Stat. 230 (codified and amended as the Richard B. Russell National School Lunch Act at 42 U.S.C. § 1751–1769j).

menus. For example, according to AMS officials, during fiscal year 2016, the agency purchased more than 110 million pounds of raw beef, over 90 percent of which was delivered to the National School Lunch Program. Further, according to AMS officials, about 41 million pounds (37 percent) were delivered raw while the rest was delivered to a federally inspected processing facility for cooking prior to delivery to school lunch program agencies.

Beef to be delivered raw to the National School Lunch program is tested for pathogens (*Salmonella* and Shiga-toxin-producing *E. coli*, two pathogens that can cause foodborne illness in humans) and certain microorganisms such as aerobic plate count bacteria, coliform bacteria, and generic *E. coli* that serve as indicators of the effectiveness of slaughter and processing plants' process controls to limit pathogens.

National School Lunch Program

According to USDA, the National School Lunch Program is a federally assisted meal program operating in public and nonprofit private schools and residential childcare institutions. It provides nutritionally balanced, low-cost or free lunches to children each school day. The program was established under the National School Lunch Act, signed by President Harry Truman in 1946. USDA's Food and Nutrition Service administers the program at the federal level. At the state level, the program is administered by state agencies, operating through agreements with school food authorities. Participating school districts and independent schools receive cash subsidies and food. In exchange, participating institutions must serve lunches that meet federal nutrition requirements and offer the lunches at a free or reduced price to eligible children. USDA's Agricultural Marketing Service purchases beef and other food for various federal nutrition assistance programs, including the National School Lunch Program.

Source: GAO analysis of U.S. Department of Agriculture (USDA) documents. Image: Courtesy of USDA. | GAO-18-272.

According to AMS officials, these indicator microorganisms indicate the quality of the food safety controls at the plant. For raw beef products that AMS considers for purchase for its programs, the agency rejects any beef that tests positive for *Salmonella*, a pathogen that can cause foodborne illness in humans. According to AMS officials, this requirement that beef purchased for these programs not test positive for *Salmonella* differs from the regulatory standard for beef inspected by USDA's Food Safety and Inspection Service (FSIS).[58] Further, according to AMS officials, AMS set this requirement because raw beef was considered the product with the most risk for recipients and enough plants were able to meet the requirement. AMS officials said that as a purchaser for various federal nutrition assistance programs, the agency has discretion to set requirements for qualified suppliers, and plants can choose whether to become qualified suppliers.

APPENDIX II: COMMENTS FROM THE U.S. DEPARTMENT OF AGRICULTURE

USDA

United States Department of Agriculture

Food Safety and Inspection Service

1400 Independence Avenue, SW, Washington, D.C. 20250

Steve D. Morris
Director
Natural Resources and Environment
United States Government Accountability Office
441 G Street, N.W.
Washington, DC

FEB 16 2018

Dear Mr. Morris,

The United States Department of Agriculture (USDA) appreciates the opportunity to review the U.S. Government Accountability Office's (GAO) draft report entitled *Food Safety: USDA Should Take Further Action to Reduce Pathogens in Meat and Poultry Products* (GAO-18-272). We also wish to acknowledge the audit team's professionalism and focus on ensuring the facts in this report were accurate and the findings germane.

General Comments

FSIS has a few general comments regarding your findings and characterization of FSIS' programs and addresses these below. We also provide our planned corrective actions for each of the recommendations for executive action. We further provide several technical edits to ensure the document is factually correct.

58 FSIS has set the maximum acceptable percentage of raw ground beef sampled to test positive for *Salmonella* at 7.5 percent.

Performance standards are one piece of the overall food safety approach to mitigate the risk of foodborne illness. There are many steps that FSIS takes to ensure the safety of the product including verifying that establishments are controlling pathogens through their Hazard Analysis and Critical Control Point (HACCP) systems, providing guidance for industry and consumers, and requiring that product be labeled with safe handling instructions for consumers. FSIS uses pathogen reduction performance standards to assess process control of establishments that prepare meat and poultry products.

GAO asserts on its Highlights page and on pages 15 and 30 that FSIS needs to clearly document its process for deciding which products to consider for new pathogen standards and, that until this process is clearly documented, the Agency will not have assurance its decisions are risk-based and Agency personnel will know the process when making such decisions. Although FSIS agrees that it can take additional steps to internally document its process for creating and updating pathogen reduction performance standards, it does not agree with GAO that the Agency does not have assurance its decisions are risk based and that Agency personnel will not know the process when making decisions. FSIS has consistently documented and published its process within the *Federal Register* through which FSIS announces or proposes new standards.[1] Agency personnel use these *Federal Register* notices (FRN) as guidance and as historical reference when developing new standards. In addition, in *Federal Register* 81 FR 7285; February 11, 2016,[2] FSIS indicates that it designed its new performance standards for raw chicken parts and comminuted chicken and turkey to achieve at least a 30 percent reduction in illnesses from *Salmonella*, and a 33 percent reduction in illnesses from *Campylobacter* in line with the reduction goals in Healthy People 2020. This FRN demonstrates how FSIS uses risk to drive policy decisions and make changes to its performance standards.

We include a few important technical comments in addition to those previously provided to GAO:

- GAO leaves out two key points in explaining what establishments are required to do under HACCP. In footnote 12 on page 3, FSIS requests GAO add language clarifying that the establishment also needs to validate the steps to prevent, eliminate, or reduce each hazard will be effective, and then it needs to verify that its food safety system is working as intended on an ongoing basis.[3] Page 9 also omits that FSIS inspectors at slaughter and processing establishments observe operations as part of their inspection tasks, in addition to GAO's statement regarding checking records to verify an establishment's compliance with its HACCP plans.
- On GAO's Highlights page and page 14, GAO states that FSIS has not developed performance standards for "widely available" products, such as turkey breasts. However, as FSIS states in 80 FR 3940, "about 85 percent of poultry products available to consumers are chicken," and the "amount of chicken parts available from fabricated broiler carcasses is larger than that of turkey carcasses that are fabricated into raw turkey parts and available to consumers." Furthermore, it states, "there is more contamination of broiler carcasses with *Salmonella* and *Campylobacter* compared to turkey carcasses. For example, in 2008, FSIS found that broiler carcasses had a *Salmonella* prevalence of 7.5 percent, while in 2009 turkey carcasses had *Salmonella* prevalence of 1.7 percent." Given the lower exposure from/to turkey parts, and the lower hazard from turkeys in general, FSIS made a risk-based decision

[1]Please see (1) "Pathogen Reduction; Hazard Analysis and Critical Control Point (HACCP) Systems" final rule, (61 FR 38806; July 25, 1996), available on the FSIS website at https://www.fsis.usda.gov/wps/wcm/connect/e113b15a-837c-46af-8303-73f7c11fb666/93-016F.pdf?MOD=AJPERES. See page 38835, section: Role of Microbiological Performance Criteria and Standards in FSIS Food Safety Strategy. (2) "New Performance Standards for *Salmonella* and *Campylobacter* in Young Chicken and Turkey Slaughter Establishments; New Compliance Guides," (75 FR 27288; May 14, 2010), available at: https://www.gpo.gov/fdsys/pkg/FR-2010-05-14/pdf/2010-11545.pdf. See page 27289, section: Salmonella Performance Standards. (3) "Changes to the *Salmonella* and *Campylobacter* Verification Testing Program: Proposed Performance Standards for *Salmonella* and *Campylobacter* in Not-Ready-to-Eat Comminuted Chicken and Turkey Products and Raw Chicken Parts and Related Agency Verification Procedures and Other Changes to Agency Sampling," (80 FR 3940; January 26, 2015), and available on the FSIS website at https://www.fsis.usda.gov/wps/wcm/connect/55a6586e-d7d6-406a-b2b9-c5d83c110511/2014-0023.pdf?MOD=AJPERES. See page 3940, section: Background.

[2]The "New Performance Standards for Salmonella and Campylobacter in Not-Ready-to-Eat Comminuted Chicken and Turkey Products and Raw Chicken Parts and Changes to Related Agency Verification Procedures: Response to Comments and Announcement of Implementation Schedule," was published in the Federal Register (81 FR 7285; February 11, 2016), is available on the FSIS website at https://www.fsis.usda.gov/wps/wcm/connect/b90ac243-af49-43a9-9fb2-00aad090345b/2014-0023.pdf?MOD=AJPERES.

[3]See 80 FR 27557, and https://www.fsis.usda.gov/wps/wcm/connect/3ba826cc-6c79-4f17-85fc-29200f4e8d05/2009-0019-2015.pdf?MOD=AJPERES.

to not include turkey parts in its baseline study and not to develop performance standards for turkey parts.

- On pages 18, 27, and 28, GAO refers to the Agency not having mandatory recall authority as a challenge to reducing *Salmonella* and *Campylobacter* contamination in poultry. However, *Salmonella* and *Campylobacter* are not ordinarily adulterants in raw product and FSIS does not consider a recall if the product is not adulterated (e.g., raw product with *Salmonella* or *Campylobacter* under normal circumstances). If the product is adulterated, FSIS requests a recall. As noted on page 27 of the report, if an establishment refuses to recall the product, FSIS has authority to withdraw inspectors, and seize or detain products, thus preventing the establishment's products from entering commerce, which effectively protects human health. Consequently, mandatory recall authority is not a challenge in protecting public health.
- Page 29 of the report references FSIS testing for *Salmonella* in certain components of ground beef. However, the report omits that FSIS also analyzes for *Salmonella* in ground beef, and beef manufacturing trimmings, and other raw ground beef components. This new approach allows for better prevalence estimates and for monitoring changes in prevalence over time.

USDA Responses to GAO Recommendations for Executive Action

GAO Recommendation 1:
The Administrator of FSIS should document the agency's process for deciding which products to consider for new standards, including the basis on which such decisions should be made.

USDA Response:
FSIS concurs with this recommendation. FSIS will complete an additional internal document that delineates the Agency's process for creating or updating performance standards and its basis for decision-making.

GAO Recommendation 2:
The Administrator of FSIS should set "timeframes" for determining what pathogen standards or additional policies are needed to address pathogens in beef carcasses, ground beef, pork cuts, and ground pork.

USDA Response:
FSIS concurs with this recommendation specific to "timeframes" within its control. In 2014, FSIS began analyzing all raw beef samples it collects for Shiga-toxin producing *E. coli* (STEC) analysis for *Salmonella* (79 FR 32436). In 2018, FSIS will continue to assess the data from these sampling projects, along with baseline data and outbreak/illness data, to help assess the risks. FSIS will use this data to determine whether new or revised standards or additional policies (e.g., training, guidance to industry, or instructions to field personnel) are needed to address *Salmonella* in beef products.

In June 2017, FSIS began the second phase of the Raw Pork Products Exploratory Study; Phase 2 will continue for approximately one year (May 2018). FSIS will analyze the data collected in the Raw Pork Products Exploratory Sampling Program, along with baseline data and outbreak/illness data, to help assess the risks from STEC in pork. In 2019, the Agency will use this data to determine whether standards or additional policies (e.g., training, guidance to industry, or instructions to field personnel) are needed to address *Salmonella* in pork products.

GAO Recommendation 3:
The Administrator of FSIS should include available information on the effectiveness of on-farm practices to reduce the level of pathogens as it finalizes its guidance for controlling *Salmonella* in hogs.

USDA Response:
FSIS concurs with this recommendation. FSIS does not have on-farm (pre-harvest) jurisdiction, yet it will incorporate available pre-harvest related research results from the published literature into its compliance guidelines. FSIS is currently revising its draft compliance guidelines to address the reduction of *Salmonella* in market hogs, and has encouraged establishments to use the on-farm component in the draft guidelines while the guidelines have been out for comment. The Agency will include available scientific information on the effectiveness of each recommended farm practice to reduce *Salmonella* in market hogs in those guidelines.

Thank you for the opportunity to review and comment on this draft report.

Sincerely,

Carmen Rottenberg
Acting Deputy Under Secretary
Office of Food Safety

In: Food Safety
Editor: Georges Martin
ISBN: 978-1-53616-762-7

Chapter 5

FOODBORNE ILLNESSES AND OUTBREAKS FROM FRESH PRODUCE*

Renée Johnson

The produce industry continues to be associated with a series of foodborne illness outbreaks across multiple U.S. states and Canada, resulting in hundreds of illnesses and hospitalizations, as well as kidney failure and death in some cases. Many in Congress have expressed concern that foodborne illness outbreaks are continuing to occur despite ongoing regulatory oversight by the Food and Drug Administration (FDA) and state public health authorities and given the enhanced authorities and resources provided to food safety officials following the enactment of comprehensive food safety legislation as part of the FDA Food Safety Modernization Act in 2010 (FSMA, P.L. 111- 353). FSMA req;lllloluired the development and implementation of produce safety standards for certain fruits and vegetables, among other regulatory actions.

* This is an edited, reformatted and augmented version of Congressional Research Service Publication No. IF11092, dated February 4, 2019.

FOODBORNE ILLNESSES LINKED TO PRODUCE

The Centers for Disease Control and Prevention (CDC) defines a "foodborne disease outbreak" as occurring when two or more people get the same illness from the same contaminated food or drink. Based on previous CDC outbreak investigations, microbial hazards associated with fruits and vegetables include pathogenic (disease-causing) strains of Shiga toxin-producing *E. coli* (STEC), *Salmonella*, Norovirus or Norwalk-like virus, and *Listeria monocytogenes*. Other produce-related hazards have involved *Vibrio*, *Shigella*, *Cryptosporidium*, *Giardia*, *Cyclospora*, *Toxoplasma gondii*, and Hepatitis A virus.

Microbial hazards may be introduced through agricultural and processing water (used in production or transported via runoff), soil amendments (such as manure and municipal biosolids), worker hygiene, field and packing facility sanitation, and produce transportation and distribution.

During the last three years (2016-2018), FDA, CDC, and state and local officials have investigated several multistate outbreaks involving leafy greens, salad mix, sprouts, cucumbers, papaya, frozen strawberries, prepared fresh and frozen vegetables, pre-cut melons, pistachios, and dried coconut. Norovirus, *Salmonella*, and *E. coli* are among the leading cause of illnesses linked to produce. Multiple foodborne illness and outbreak investigations involved *E. coli* O157:H7 illnesses that were linked to leafy greens. Of these, one was linked to romaine lettuce grown in California, and another was linked to romaine lettuce grown in Arizona. Other investigations involved *E. coli* linked to leafy greens (not traced back to a common supplier) and involved *Listeria monocytogenes* linked to packaged salad.

Figure 1 and *Figure 2* show reported outbreak and illness trends for all produce and a subset of produce for leafy greens based on data from the CDC's National Outbreak Reporting System (NORS). These data indicate the number of single and multistate outbreaks associated with produce ranged from about 30 to 60 per year and sickened from 900 to nearly 3,000 people annually (1998-2016) (*Figure 1*).

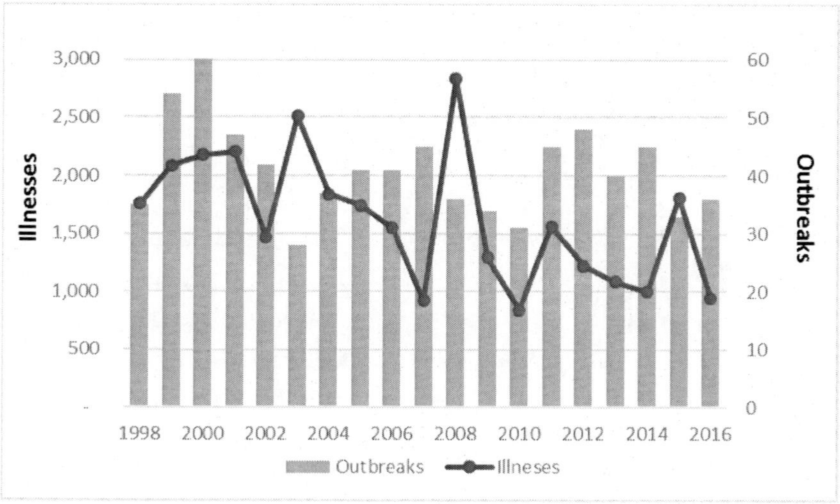

Figure 1. Foodborne Illness Outbreaks, All Produce.

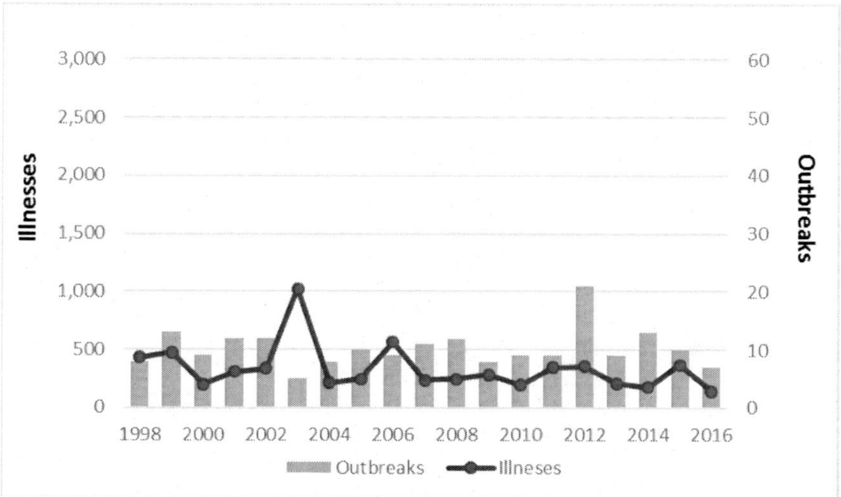

Source: CRS from NORS data, https://www.cdc.gov/nors/index.html. Produce includes fruits, nuts (excluding peanuts), vegetables, sprouts, root crops, and herbs. Leafy greens is a subset of all produce and includes all lettuce, greens, and cabbage (excluding sprouts).

Notes: Includes confirmed and suspected cases across reported single and multistate outbreaks and all attributed causes and settings (restaurants and institutions, e.g., schools, hospitals, and prisons).

Figure 2. Foodborne Illness Outbreaks, Leafy Greens.

Leafy greens, a subset of the produce category, accounted for between 20% and 40% of these reported outbreaks and 10% to 40% of all produce-related illnesses (1998-2016). Over this period, single and multistate outbreaks associated with leafy greens ranged from about 10 to 20 per year and sickened between 100 and 1,100 annually (*Figure 2*).

Analysis by the Center for Science in the Public Interest, a public health advocacy group, indicates that foodborne illness outbreaks associated with produce were more common and sickened more people than any other food category. A 2008 United Nations study ranks leafy greens as the "highest priority" in terms of microbiological hazards within the fruit and vegetable category. This ranking is based on the sheer number of outbreaks and the types of microbial hazard associated with leafy greens, but it also reflects the size and scale of production and trade in leafy greens. Other factors include the diversity and complexity of production of leafy greens and potential for amplification of foodborne pathogens through the food chain due to certain post-harvest activities (e.g., whether packaged in- field or pre-cut and bagged). Available information tracked by CDC as part of its Foodborne Diseases Active Surveillance Network indicate that, across all food groups, infections caused by certain STEC infections have been decreasing, while the incidence of *Salmonella* infections has not changed significantly. Making such generalizations, however, is often complicated by differences in laboratory testing methods and/or differences depending on the strain of a microorganism.

FDA Standards for Produce Safety

FSMA amends the Federal Food, Drug, and Cosmetic Act (21 U.S.C. §§301 *et seq.*), which governs foods under FDA's jurisdiction. Under FSMA, FDA has developed and implemented mandatory food safety and traceability requirements for farmers, packers, and processors of domestically produced and imported products. Selected provisions that broadly address produce are shown in the *text box*.

Selected FSMA Provisions Related to Produce

Inspections of Records (§101)

- Allows FDA to inspect records related to the "manufacture, processing, packing, distribution, receipt, holding, or importation" of certain food and feed.

Registration of Food Facilities (§102)

- Requires that food facilities be subject to biennial registration renewal. Provides that FDA may suspend a facility's registration in certain cases.

Hazard Analysis and Risk-Based Preventive Controls (§103)

- Requires FDA to establish mandatory preventive controls for food facilities except for "small" and "very small" businesses.

Standards for Produce Safety (§105)

- Requires FDA to establish mandatory minimum standards for the safe production and harvesting of fruits and vegetables except for "small" and "very small" businesses.

Targeting of Inspection Resources (§201)

- Requires FDA to identify high-risk facilities, increase the frequency of inspection of domestic and foreign facilities, identify and conduct inspections at ports of entry, and improve inter-agency coordination and cooperation.

Tracking and Tracing Food, Records (§204)

- Requires FDA to establish pilot projects to improve traceability of foods and to establish additional recordkeeping requirements for certain "high-risk foods."

Surveillance (§205)

- Requires CDC to enhance foodborne illness surveillance systems and to conduct an assessment of state and local food safety and defense capacities.

At the farm production level, FSMA principally affects produce growers by directing FDA to establish produce safety standards (§105). FDA finalized its produce safety regulation in 2015 (80 *Federal Register* 74353). FDA's produce rule addresses certain identified routes of potential contamination, covering water and soil amendments used in production, domesticated and wild animal intrusions in production area, worker training and hygiene, and equipment and sanitation practices used in production. The regulation covers fruits and vegetables (including mixes), mushrooms, sprouts, peanuts, tree nuts, and herbs. FDA estimates that the regulation covers as many as 40,000 domestic produce farms and nearly 15,000 foreign farms. Foods not covered by regulation include foods that are rarely consumed raw (listed at 21 CFR 112.2(a)(1)), foods that go to commercial processing, foods produced for personal consumption, and certain foods identified as low risk. Produce that undergoes certain commercial processing, such as bagged salads and leafy greens, are further covered by FDA's rule on preventive controls affecting food facilities (§103). Other FSMA requirements affecting on- farm activities include facility registration, inspection, and records access. Certain qualified farms and facilities are exempt from regulation depending on business size, among other factors.

Deadlines for complying with FDA's produce rule are being phased in over time and vary depending on the crop and farm size. In some cases, FDA has granted more time to some farms to implement and comply with certain requirements. For example, in September 2017, FDA proposed to extend the compliance deadline for agricultural water testing because of concerns about the feasibility and cost involved (82 *Federal Register* 42963). Under the extension, larger farms have until 2022 to comply with the rule's water requirements, while small and very small farms have until 2023 and 2024, respectively, to comply.

FSMA also addresses food traceability (§204). Traceability refers to the ability to fully trace the movement of food and ingredients through each specific stage of production, processing, and distribution and to be able to identify the origin of food and ingredients when a food or finished product is found to be unsafe. Full traceability often requires extensive recordkeeping and/or other types of traceback mechanisms. FSMA required FDA to establish pilot projects and improve its capacity to effectively and rapidly track and trace foods in the event of an outbreak.

FSMA also required FDA to designate high-risk foods that require additional recordkeeping to protect public health. FDA's pilot projects were completed in 2012. However, FDA has not yet identified or established enhanced recordkeeping requirements for high-risk foods.

CONSIDERATIONS FOR CONGRESS

As foodborne illness outbreaks continue to occur, some continue to question the effectiveness of FDA's food safety regime. Regarding lapses in produce safety, some advocacy groups point to FDA's repeated delays in fully implementing key FSMA produce standards, including postponing compliance with the rule's water quality testing requirements, as a contributing factor. Some groups further blame FDA for not fully implementing FSMA's traceability requirements, especially regarding high-risk foods. Critics also continue to cite the agency's low rates of inspection and enforcement of FDA-regulated farms and food facilities. Others claim that FDA has not been provided with the resources to hire the additional inspectors that are needed to fully implement and adequately enforce the produce standards. Still others point to a possible lack of coordination between FDA and state and local authorities, which often bear most of the responsibility for inspecting farms and food facilities within their jurisdictions.

In: Food Safety
Editor: Georges Martin

ISBN: 978-1-53616-762-7

Chapter 6

PROPOSED REORGANIZATION OF U.S. FEDERAL FOOD SAFETY AGENCIES*

Renée Johnson and Agata Dabrowska

The Trump Administration is proposing to consolidate the federal government's primary food safety functions into a single federal agency based in the U.S. Department of Agriculture (USDA). This proposal is part of its broader effort to reorganize the U.S. federal government under its reform plan and reorganization proposals released in June 2018. Some in Congress—and also previous Administrations—have long pursued similar efforts to consolidate U.S. federal food safety functions.

ORGANIZATION OF THE FOOD SAFETY SYSTEM

Numerous federal, state, and local agencies share responsibilities for regulating the safety of the U.S. food supply. Federal responsibility for food safety rests primarily with USDA and the Food and Drug

* This is an edited, reformatted and augmented version of Congressional Research Service Publication No. IF10974, dated November 8, 2017.

Administration (FDA). FDA, an agency of the Department of Health and Human Services (HHS), is responsible for ensuring the safety of all domestic and imported foods except meat and poultry.

USDA's Food Safety and Inspection Service (FSIS) regulates meat, poultry, some egg products, and catfish.

At FDA, the Center for Food Safety and Applied Nutrition (CFSAN) oversees the safety of food and cosmetic products. CFSAN's primary responsibilities include the safety of most foods, food additives, infant formulas, medical foods, dietary supplements, and foods and ingredients developed through biotechnology; programs addressing health risks associated with foodborne, chemical, and biological contaminants; and food and nutrition labeling. CFSAN also collaborates with FDA's Office of Regulatory Affairs to conduct field activities, including facility inspections. FDA's authority extends beyond foods to also include pharmaceutical drugs, medical devices, biologics, animal drugs and feeds, veterinary devices, radiation-emitting products, and tobacco.

At USDA, FSIS conducts continual inspection at federally inspected facilities that slaughter meat and poultry. In these inspections, FSIS ensures that state inspection standards are at least equivalent to federal standards and that imported meat and poultry products are produced under standards equivalent to U.S. inspection standards. Other USDA agencies also play a role in U.S. food safety oversight. For example, USDA's Animal and Plant Health Inspection Service oversees animal and plant health, including preventing the introduction of foreign diseases and pests.

USDA's Agricultural Marketing Service establishes quality and marketing grades and standards for a range of agricultural products. USDA's research agencies are also involved in food safety, providing federal funding and collaborating with universities and research institutions.

The first *text box* summarizes the primary U.S. food safety laws governing FDA and FSIS. The laws listed are among roughly 30 related to food safety, according to Government Accountability Office (GAO) estimates.

GAO has also identified up to 15 federal agencies, including FDA and FSIS, as collectively responsible for ensuring the safety of food produced and sold in the United States. State and local food safety authorities collaborate with federal agencies on inspection and other food safety functions.

Primary U.S. Food Safety Laws

HHS's Food and Drug Administration (FDA)

- Federal Food, Drug, and Cosmetic Act (FFDCA, 21 U.S.C.
- §§341 *et seq.*), as amended by the FDA Food Safety Modernization Act (FSMA, 21 U.S.C. §§2201 *et seq.*)
- Fair Packaging and Labeling Act (15 U.S.C. §1454)
- Public Health Service Act (42 U.S.C. §§201 *et seq.*)
- Federal Import Milk Act (21 U.S.C. §§141-149)
- Federal Anti-Tampering Act (18 U.S.C. §1365)
- Pesticide Monitoring Improvements Act (21 U.S.C. §1401)

USDA's Food Safety and Inspection Service (FSIS)

- Federal Meat Inspection Act (FMIA, 21 U.S.C. §§601 *et seq.*)
- Poultry Products Inspection Act (PPIA, 21 U.S.C. §§451 *et seq.*)
- Egg Products Inspection Act (EPIA, 21 U.S.C. §§1031 *et seq.*)
- Agriculture Marketing Act (7 U.S.C. §§1621 *et seq.*)

Source: CRS Report RS22600, *The Federal Food Safety System: A Primer.*

PROPOSED FOOD SAFETY REORGANIZATION

The Trump Administration's proposal to reorganize the U.S. federal government includes a proposal to consolidate FSIS and FDA's food safety functions into a single "Federal Food Safety Agency" to be housed at USDA. According to the Administration, this effort would address a "fragmented and illogical division of federal oversight" and would "merge approximately 5,000 full-time equivalent (FTE) employees and $1.3 billion from FDA with about 9,200 FTEs and $1 billion in resources in USDA." FDA would be renamed the "Federal Drug Administration" and its focus would be on drugs, devices, biologics, tobacco, dietary supplements, and cosmetics.

To the extent a reorganization would transfer an agency or entity vested by law in a particular department to a different department, additional legislation might be needed. These food safety laws specifically delegate authority to HHS in the case of FFDCA and FSMA and to USDA in the case of the FMIA, PPIA, and EPIA.

PREVIOUS REORGANIZATION EFFORTS

The organization of the U.S. food safety system has been debated on and off since FDA was removed from USDA in the 1940s. Since then, a number of congressional and executive branch initiatives have raised the prospect of creating a single federal food safety agency. For example, in 1949, a presidential commission under the Truman Administration proposed transferring federal food safety activities to USDA. Years later, Congress passed the Wholesome Meat Act of 1967 and the Wholesome Poultry Products Act of 1968, substantially restructuring U.S. meat and poultry products inspection. Also during this time, a 1969 *White House Conference on Food, Nutrition and Health* report highlighted the divergence in food safety policy between USDA and the Department of Health, Education, and Welfare (which later became HHS), where FDA resided at that time. In the following years, there were a series of reports and congressional hearings adding to the debate over a single food safety agency. For example, in 1977, the Senate Committee on Governmental Affairs conducted its *Study on Federal Regulation*, recommending that USDA's food safety functions be transferred to FDA.

Committee Jurisdiction for Food Safety Issues

House: The Committee on Energy and Commerce claims general jurisdiction over all FDA-regulated products, including foods. The Committee on Agriculture asserts jurisdiction over USDA's inspection programs.

Senate: The Committee on Health, Education, Labor, and Pensions claims jurisdiction over all FDA-regulated products, including foods. The Committee on Agriculture, Nutrition, and Forestry asserts jurisdiction over USDA's inspection programs.

Congressional Appropriations Committees: The Agriculture subcommittees are responsible for funding and oversight of both FDA and USDA in the House and Senate.

Recent Congressional Efforts

Some in Congress have advocated for reforms to the nation's food safety system, particularly with respect to coordination and organization among federal agencies. Efforts to establish a single food safety agency were active from the 103rd Congress through the 111th Congress.

Although Congress passed comprehensive food safety legislation in 2010 (FSMA, P.L. 111-353)—representing the largest expansion and overhaul of FDA's food safety authorities since the 1930s—FSMA did not alter the existing food safety jurisdiction between FDA and USDA. While FSMA was being debated, however, Congress did consider options to reorganize and consolidate certain federal food safety functions. For example, in the 111th Congress, H.R. 875 (Food Safety Modernization Act of 2009) included provisions to consolidate certain food safety functions into a single federal agency, the Food Safety Administration, to be housed within HHS.

Following FSMA's enactment, some in Congress continued to be interested in establishing a single federal food safety agency. In the 114th Congress, the Safe Food Act of 2015 (H.R. 609/S. 287) proposed creating a single independent Food Safety Administration, which would be responsible for regulating food safety and related labeling, inspection, enforcement, and research functions of both domestically produced and imported foods. H.R. 609/S. 287 proposed transferring and consolidating the food safety authorities at FDA and USDA as well as portions of the National Marine Fisheries Service at Department of Commerce.

Recent Administration Efforts

The Obama Administration proposed to establish a single federal food agency as part of its FY2016 budget request, claiming that a single food safety agency would "provide focused, centralized leadership, a primary voice on food safety standards, and clear lines of responsibility and accountability that will enhance both prevention of and responses to outbreaks of foodborne illnesses." The proposal would not have created a new independent agency but would have instead transferred existing food safety functions into a new agency within HHS. The proposal was not included in the enacted appropriation.

Views Expressed by Outside Groups

Establishing a single food agency has the support of GAO and the National Academies of Sciences, Engineering, and Medicine and others within academia, as documented in myriad studies and reports. GAO's reports on food safety have regularly highlighted that the U.S. food safety system is fragmented across different departments and agencies that have overlapping responsibilities and that a single food safety agency would improve effectiveness and efficiency of food safety regulation.

While some view consolidation as a way to improve efficiency and effectiveness, others worry that it could unnecessarily compromise day-to-day food safety efforts. Some food safety advocates claim that USDA's principal purpose is supporting agribusiness and that food safety regulation and promotion should remain separate. Some also raise concerns about merging two agencies with distinctly different authorities, inspection regimes, and cultures. Others say it is unclear what this would mean for ongoing FSMA implementation at FDA and for both domestic and international food producers affected by the law's regulations. Others assert that any federal overhaul would be complex and costly, given the sheer number of federal agencies and different statutes involved, and could create new forms of fragmentation. Some also anticipate resistance by congressional committees that cover different agency jurisdiction, powers, and expertise. (Authorizing committees are identified in the second text box.)

Most consumer advocacy groups have opposed consolidating food safety operations within HHS, claiming that HHS does not have the necessary expertise or adequate resources to manage an expanded food safety function.

Others point to differing legal authorities and inspection approaches. Others have claimed that USDA has a better record regarding food safety inspection and enforcement and worry that transferring these functions to HHS will lower meat and poultry inspection standards.

Creating a single food safety agency housed at USDA often raises concerns about the complexity of FDA's oversight authorities, which cover not only foods and beverages but other food-related responsibilities (e.g., safety of additives, biotechnology) and products. For example, under current law, dietary supplements are regulated as part of FDA's food safety responsibilities. The Trump Administration's proposal would keep supplements within the purview of the renamed "Federal Drug Administration," which is opposed by many within the industry.

INDEX

D

E

F

S

T

U

V

W